Reliability Models for Engineers and Scientists

Reliability Models for Engineers and Scientists

Mark P. Kaminskiy

CRC Press
Taylor & Francis Group
Boca Raton   London   New York

CRC Press is an imprint of the
Taylor & Francis Group, an **informa** business

CRC Press
Taylor & Francis Group
6000 Broken Sound Parkway NW, Suite 300
Boca Raton, FL 33487-2742

First issued in paperback 2017

© 2013 by Taylor & Francis Group, LLC
CRC Press is an imprint of Taylor & Francis Group, an Informa business

No claim to original U.S. Government works
Version Date: 2012920

ISBN 13: 978-1-138-07756-0 (pbk)
ISBN 13: 978-1-4665-6592-0 (hbk)

Library of Congress Cataloging-in-Publication Data

Kaminskiy, Mark, 1946-
    Reliability models for engineers and scientists / Mark P. Kaminskiy.
      p. cm.
    Includes bibliographical references and index.
    ISBN 978-1-4665-6592-0 (hardback)
    1. Reliability (Engineering)--Mathematics. 2. Science--Data processing. I. Title.

TA169.K36 2012
620'.00452--dc23   2012031641

Visit the Taylor & Francis Web site at
http://www.taylorandfrancis.com

and the CRC Press Web site at
http://www.crcpress.com

Contents

Introduction

The main purpose of this book is to make modern mathematical reliability models available to reliability engineers and risk analysts, as well as to physicists (especially those working in the physics of failure), biologists, gerontologists, and many others. The intention is to represent these models in simple terms, discussing in detail their physical meaning and the real data supporting (or not) the adequacy of the models.

For more than 40 years, the Barlow and Proschan monograph (1965) has been the unprecedented classical book on mathematical reliability models—and it still is. It should be noted that this and another later monograph (Barlow and Proschan, 1975) are the only exceptionally probabilistic (as opposed to *statistical*) resources on reliability. From a variety of standpoints, the merits of these resources cannot be overestimated, including from the standpoint of introductory reading, because they are not loaded with the respective statistical analysis procedures.

Why do we need a new book? In recent years, many new results related to mathematical reliability theory have appeared. In the latest edition of the Barlow and Proschan monograph published in SIAM's Classics in Applied Mathematics series (Barlow and Proschan, 1996), it is stated that "much has transpired since its original publication in 1965." The book offered here to the reader is intended to discuss in simple terms the basic reliability notions and models, including some new results to the extent that they might be useful for practitioners.

Notably, some of the real data examples the reader will find in this book might look quite old. They reveal the fact that sometimes there are few (if any) real data sets available to support a given model. On the other hand, this shows that the number of reliability concepts and models is often bigger than the number of the respective supporting data sets. One can also hope that an understanding of the situation will encourage readers to publish more experimental data, which are so desperately needed to further progress in reliability theory and practice. In order to provide a sense of the difference between the real and ideal data, many simulated data sets are given.

Traditionally in reliability books, each probabilistic model is followed by the respective statistical procedures, depending on the type of available data. In order to get a better understanding of probabilistic reliability models and their "supportability" by real data, a statistical data analysis is not considered in this book. Instead, the reader is provided with some basic references to reliability data analysis in books or in papers. It should be noted that many comprehensive books on statistical reliability/survival data analysis are available: Nelson (1981, 1990, 2003, 2004), Lawless (1982, 2002), Rigdon and Basu (2000), and Meeker and Escobar (1998), to name just a few.

The book consists of an introductory chapter (1) and two main chapters (2 and 3). In Chapter 2, the time-to-failure distributions for the nonrepairable objects are introduced as arising from some simple random shocks (damage) models. In Chapter 3, the failure process models are introduced by either making some assumptions about the underlying (time between failures) distributions or, vice versa, by making some assumptions about a failure process, from which one could arrive at the respective underlying distribution.

The book introduces a new concept of the Gini-type index. Applied to aging/rejuvenating components (nonrepairable systems) in Chapter 2, the index is a measure showing how different a given aging/rejuvenating distribution is compared to the exponential distribution. In Chapter 3, a similar index is applied to aging/rejuvenating repairable systems, performing a kind of bridge between these two chapters.

This resource should be suitable for students of reliability engineering as well as for those who wish for a supplement on applied survival data analysis.

I am most grateful to my long-term colleague and coauthor Dr. Vasiliy Krivtsov of the Ford Motor Company for numerous useful discussions about the book. Some sections of the book are based on our mutual publications. I also acknowledge the editorial help of Professor Leonard Rinchiuso of West Liberty University.

Mark Kaminskiy
NASA Goddard Space Flight Center, ManTech International Corporation
Greenbelt, Maryland

About the Author

Mark Kaminskiy is currently a principal reliability engineer at the NASA Goddard Space Flight Center (via ManTech International). In the past, he conducted numerous research and consulting projects funded by government and industrial companies such as the Department of Transportation, the Coast Guard, the Army Corps of Engineers, the Department of Homeland Security, the Navy, the Nuclear Regulatory Commission, General Dynamics, the American Society of Mechanical Engineers, Ford Motor, Qualcomm, Inc., and several other engineering companies. He has taught several graduate courses on reliability engineering at the University of Maryland. Dr. Kaminskiy is the author or coauthor of over 50 professional publications, including a book on reliability engineering and risk analysis, and chapters in many books on statistical reliability and risk analysis. He received his MS degree in nuclear physics from the Technical University of St. Petersburg (Russia) and PhD in electrical engineering from the Electrotechnical University of St. Petersburg (Russia).

1

Time-to-Failure Distributions and Reliability Measures

> One needs to know one's measure when drinking vodka.
>
> —**A Russian proverb**

> Every expert is like a gumboil: his fullness is one-sided.
>
> —**Kozma Prutkov**

This chapter is about some basic definitions and probabilistic characteristics, which are used throughout the entire book. The reliability measures discussed below are used mostly for nonrepairable objects (components). The measures related to repairable objects (systems) are discussed later, in Chapter 3.

1.1 Probability Density and Cumulative Distribution Functions

We begin with the most common variable in reliability: the *continuous* positively defined random variable—*time to failure* (TTF), τ.

Let's have a nonrepairable object (system, component). If the object's TTF has a probability density function (PDF) $f(\tau)$, the probability $F(t)$ that a failure occurs during the time interval $[0, t]$ is given by the following integral:

$$F(t) = \int_0^t f(\tau)d\tau \tag{1.1}$$

Depending on context, the time t is referred to as the *age* (counted from $t = 0$), *mission duration*, or *exposure*. Note that the exposure can be not only time, but miles, revolutions, cycles, or, say, electric field strength, e.g., if one deals with dielectric strength.

In probability theory, the function $F(t)$ is known as the *cumulative distribution function* (CDF) or *unreliability function*. In reliability engineering, the same function might be referred to as *unreliability function* or *failure probability function*. In risk analysis, the function $F(t)$ might be called the *probability of mission failure*. In statistical physics, PDF $f(t)$ is known as the *distribution function*.

The function corresponding to CDF (1.1), known as the *reliability function, survival* function, *survivor* function, or simply *reliability R(t)*, is defined as

$$R(t) \equiv 1 - F(t) \tag{1.2}$$

1.2 Conditional Reliability, Failure Rate, Cumulative Failure Rate, and Average Failure Rate

If our object has survived to age t, the *conditional probability of failure (conditional CDF)* during the next time interval $(t, t + x]$, $F(x|t)$ is

$$F(x|t) = \frac{F(t+x) - F(t)}{R(t)} \tag{1.3}$$

The respective *conditional PDF h(t)*, which is called the *failure rate* (or *hazard rate* or, in actuarial science,[*] the *force of mortality*) is introduced as

$$h(t) = \lim_{x \to 0} \frac{1}{x} \frac{F(t+x) - F(t)}{R(t)} =$$

$$= \frac{1}{R(t)} \lim_{x \to 0} \frac{F(t+x) - F(t)}{x} = \frac{f(t)}{R(t)} \tag{1.4}$$

In the following, the term *failure rate* is used as the one well established in reliability today (Hoyland and Rausand, 1994, 2004). The term *failure rate* should not be confused with the term *rate of occurrence of failures* (ROCOF). Whereas the term *failure rate* is applied to nonrepairable (nonrenewable) objects only, ROCOF is exclusively related to repairable/renewable objects. We will discuss the issue in detail when ROCOF is introduced in Chapter 3.

Integrating both sides of (1.4), the following useful relation can be obtained:

$$R(t) = \exp\left[-\int_0^t h(\tau)d\tau \right] \tag{1.5}$$

Similar to the relation between the probability density function and the cumulative distribution function (1.1), the so-called *cumulative failure rate H(t)* is introduced as

$$H(t) = \int_0^t h(\tau)d\tau \tag{1.6}$$

---

[*] Actuarial science applies probabilistic and statistical risk analysis methods to insurance problems.

TABLE 1.1

Summary of Probabilistic Functions and Their Relationships

	$f(t)$	$F(t)$	$R(t)$	$h(t)$	$H(t)$
$f(t) =$	*	$F'(t)$	$-R'(t)$	$h(t)\exp\left(-\int_0^t h(\tau)d\tau\right)$	$H'(t)\exp(-H(t))$
$F(t) =$	$\int_0^t f(\tau)d\tau$	*	$1-R(t)$	$1-\exp\left(-\int_0^t h(\tau)d\tau\right)$	$1-\exp(-H(t))$
$R(t) =$	$1-\int_0^t f(\tau)d\tau$	$1-F(t)$	*	$\exp\left(-\int_0^t h(\tau)d\tau\right)$	$\exp(-H(t))$
$h(t) =$	$\dfrac{f(t)}{\int_t^\infty f(\tau)d\tau}$	$\dfrac{F'(t)}{1-F(t)}$	$\dfrac{-R'(t)}{R(t)}$	*	$H'(t)$
$H(t) =$	$-\ln\left(\int_t^\infty f(\tau)d\tau\right)$	$-\ln(1-F(t))$	$-\ln(R(t))$	$\int_0^t h(\tau)d\tau$	*

Note: $F'(t) = d/dt[F(t)]$.

so the reliability $R(t)$ can be written as

$$R(t) = \exp[-H(t)] \tag{1.7}$$

Table 1.1 summarizes the above probabilistic functions and their relationships.

In the following, we will also need the *average failure rate*, which is simply introduced as the average of the failure rate over a given time interval, say, $[0, t]$, as

$$\overline{h(t)} = \frac{1}{t}\int_0^t h(\tau)d\tau = \frac{H(t)}{t} = \frac{1}{t}\int_0^t \frac{f(\tau)}{R(\tau)}d\tau = -\frac{\ln(R(t))}{t} \tag{1.8}$$

In Chapter 2, we will deal with the TTF distributions having increasing (decreasing) failure rates, as well as the TTF distributions having increasing (decreasing) average failure rates.

1.3 Reliability Measures

We have already introduced two reliability measures: the probability of a failure during the time interval $[0, t]$ $F(t)$, which is the CDF (1.1), and the corresponding reliability function $R(t)$ given by (1.2). We also have introduced the *conditional probability of failure* (1.3), with which the *conditional reliability* is introduced as

$$R(x|t) = \frac{R(t+x)}{R(t)} \qquad (1.9)$$

The mean time to failure (MTTF), which is also often called *mean life*, is the mean of the TTF distribution that was introduced in Section 1.1. Thus, using our PDF $f(t)$, we can write MTTF as

$$\text{MTTF} = \int_0^\infty \tau f(\tau) d\tau \qquad (1.10)$$

under the following rather mild condition:

$$\lim_{t \to \infty} tf(t) = 0,$$

which will always be satisfied in the following. The MTTF can be expressed through the respective reliability function $R(t)$ as

$$\text{MTTF} = \int_0^\infty R(\tau) d\tau \qquad (1.11)$$

Let us come back to the situation in which a nonrepairable object has survived to age t. The *mean residual TTF* (*mean residual life*) is introduced as

$$MResTTF(t) = \frac{1}{R(t)} \int_t^\infty R(\tau) d\tau, \qquad (1.12)$$

so that the mean time to failure (1.10) is a particular case of (1.12), in which $t = 0$. Note also that in contrast to MTTF, which is a constant, the *mean residual TTF* is a function of the component age t. For more information about the mean residual TTF, our reader is referred to Guess and Proschan (1988).

The *p-level quantile* of TTF with CDF $F(t)$ is defined as the time t_p, such that

$$F(t_p) = p, \qquad 0 < p < 1$$

It is clear that the quantile is the inverse function of the cumulative distribution function $F(t)$. If the fraction p is given as $p100$, i.e., in percent, the respective time is referred to as *percentile*.

From a reliability standpoint, the quantile (percentile) of TTF is the time at which reliability $R(t_p) = 1 - p$ (or reliability $R(t_{p100}) = (100 - p100)\%$). Various notations are used for the percentiles in engineering. For example, in accordance with the American Bearing Manufacturers Association Std-9-1990, the 10th percentile is called L_{10} life. Sometimes, it is called B_{10}, "B ten" bearing life (Nelson, 1990).

If the failure probability p (or reliability) is 0.5 (50%), the respective quantile (percentile) is called the *median*. Median is a very popular reliability measure. If one has to choose between the mean time to failure and the median (as the competing reliability measure), the latter might be a better choice, because the median is easier to perceive using one's common sense and because the median statistical estimation is, to some extent, more robust.

The failure rate (1.4) is used as a reliability measure when it is constant, the only case of which is the exponential distribution discussed in Chapter 2.

Another popular reliability measure is the average failure rate (1.8), which is sometimes called *equivalent failure rate*. Using the average failure rate, a time interval (exposure, mission time) should be given. In some situations, the time interval can be conveniently chosen as the respective p-level quantile t_p. In this case, the average failure rate takes on the following expressive form:

$$\overline{h(t_p)} = -\frac{\ln(1-p)}{t_p} \tag{1.13}$$

Let's consider one of the possible uses of the average failure rate (1.8) in reliability engineering. In many projects, the reliability requirements are given in terms of reliability $R(T)$, where T is the exposure (or mission duration), which can be expressed in units such as hours, cycles, or revolutions per minute (rpm). Let us assume that before the beginning of a reliability improvement program, the reliability of a highly reliable subsystem was estimated as $R(T) = 99.991\%$ at the exposure $T = 100,000$, say, hours. At the end of the program, the subsystem reliability was estimated once more. It turned out to be $R(T) = 99.999\%$, which corresponds to the reliability growth of 0.008%. At first sight, the reliability growth looks miserable, and one begins cursing those *nines*, which one has to deal with in developing very reliable products. However, using the same three numbers, i.e., 100,000 hours, 99.991%, and 99.999%, and evaluating the respective average failure rates (1.8), we get the reliability growth of 88.9%. Note that the reliability requirements expressed in terms of $R(T)$ usually do not specify the time-to-failure distribution—mathematically speaking, the requirements are *distribution-free*, or *nonparametric*. The same is true for the average failure rate—it is distribution-free as well. Of course, the same 88.9% improvement can be obtained using the CDF, i.e., as a decrease in the respective failure probability, i.e., in $(1 - R(T))$.

Exercises

1. Prove every relation from Table 1.1.

2

Probabilistic Models for Nonrepairable Objects

The pure and simple truth is rarely pure and never simple.

—Oscar Wilde

In this chapter, we introduce the most popular distributions used in reliability and risk analysis—binomial, Poisson, exponential, gamma, normal, lognormal, and Weibull. Each of these distributions is related to some simple physical model and illustrated by a case study. Then, the classes of time-to-failure distributions are introduced in a similar way. In order to evaluate the aging/rejuvenating of components (nonrepairable systems), the Gini-type index is introduced as a measure showing how different a given aging/rejuvenating distribution is from the exponential one. In the last part of the chapter, reliability models with explanatory variables—accelerated life model and proportional hazards model (with constant and time-dependent variables)—are discussed, as well as the mixture model and competing risks model. The terms *nonrepairable object* and *component* are used as synonyms.

2.1 Shock Models and Component Life Distributions

In this chapter, we are going to comply, to the extent possible, with the following plan. First, we introduce a shock model. In the framework of the model, the respective shocks are those random, perilous events that, one way or another, result in a component failure. Then, a failure is quantitatively defined in terms of the shock model. Ultimately, the model results in a random time-to-failure and the respective time-to-failure (TTF) distribution, from which all the reliability measures can be obtained.

2.1.1 Poisson Distribution and Homogeneous Poisson Process

In order to present the simplest shock model, we need to introduce the homogeneous Poisson process (HPP). And in order to bring in this *point process,* one has to introduce the Poisson distribution. In other words, the Poisson distribution and HPP are closely related to each other.

The Poisson distribution is one of the most popular distributions in modern science and engineering. However, we have to begin with something simple, like the Bernoulli trials. In order to make it even simpler, we begin with tossing an *ideal fair* coin. The word *fair* means that in each trial, the probability of getting a head is equal to the probability of getting a tail, and each of the probabilities is equal to exactly 0.5. We can skip the word *fair* for the definition of the Bernoulli trials, replacing the tails probability of 0.5 by any probability $0 > p > 1$ (say, $p = 0.3$), and respectively replacing the heads probability with the probability $q = 1 - p$ ($q = 0.7$). Tossing this *unfair* coin will still be within the definition of the Bernoulli trials. The word *ideal*, in our context, means that one can perform any large number of tosses with 100% assurance that probabilities p and q are constant during all the tosses.[*] In reliability terms, our ideal coin does not reveal any aging/deterioration.

The probabilities of getting heads and tails in the Bernoulli trials are given by the Bernoulli distribution. We skip discussing this distribution because, as we will see later, the Bernoulli distribution is a particular case of the *binomial* distribution.

The *binomial* distribution is not only one of the most important distributions in reliability; it is the most fundamental distribution in probability and statistics. Paying a tribute to the importance of this distribution, let's replace the gambling terms *head* and *tail* with more appropriate ones for reliability and risk analysis—*success* and *failure*.

The binomial distribution can be introduced as a set of probabilities of k successes (or failures) in n Bernoulli trials $\Pr(k; n, p)$, where the probability of success in each of n trials is p, and the probability of failure is $(1 - p)$. It is clear that k can take on an arbitrary integer between 0 and n, i.e., $0 \le k \le n$. This set of probabilities, called *probability mass function* or (as in the case of continuous random variables) the *probability density function*, is given by

$$\Pr(k; n, p) = \binom{n}{k} p^k (1 - p)^{n-k} \tag{2.1}$$

where

$$\binom{n}{k}$$

denotes the total number of sequences of n trials, in which k events of interest can occur without regard to the order of occurrence, and

---

[*] Please do not ask what *100% assurance* is or what *probability* is. Let's keep it as "do not ask, do not tell."

$$\binom{n}{k} = \frac{n!}{k!\,(n-k)!}$$

The above-mentioned Bernoulli distribution is the particular case of the binomial distribution with $n = 1$.

The mean of the binomially distributed number of successes k obviously is

$$E(k) = np \tag{2.2}$$

And its standard deviation (STD) is given by

$$STD(k) = \sqrt{np(1-p)} \tag{2.3}$$

The most common reliability engineering applications of the distribution are the so-called one-shot systems, e.g., automobile airbags, military torpedoes, and missiles. Another application is the k-out-of-n structure in the system reliability analysis, which is an example from reliability engineering—and the list can be continued. The applications of binomial distribution to reliability and risk analysis problems are not limited to those cases in which failure or success probabilities are assumed to be constant. As an example, consider the above-mentioned k-out-of-n structure, consisting of the identical (from a reliability standpoint) components with reliabilities $R_i(t) = R(t)$. Assuming that failures of the system components are statistically independent, the system reliability $R_{k\text{-}out\text{-}of\text{-}n}$ is given by the expression for the probability of observing not less than k successes in n binomial trials, i.e.,

$$R_{k\text{-}out\text{-}of\text{-}n}(t) = \sum_{i=k}^{n} \binom{n}{i} [R(t)]^{i} [1 - R(t)]^{n-i}$$

In the above expression, the component reliability $R(t)$ is time dependent. It can be a reliability function of any TTF distribution—exponential, Weibull, gamma, etc.

Now we are about to introduce the Poisson distribution, which is another fundamental discrete distribution in reliability and risk studies. For our purposes, the Poisson distribution is mainly used as a distribution of a number of events. However, these events might be, for example, some defects randomly located in a piece of silicon dioxide.

The Poisson distribution gives the probability of observing k events in a given interval (or, generally speaking, in a domain) as the following:

$$\Pr(k; M) = \frac{M^{k} e^{-M}}{k!} \tag{2.4}$$

where $k = 0, 1, 2, \ldots$, and M is the mean number of events, which is the only parameter of the Poisson distribution.

If the interval in which the events of interest are being observed is specified, say, as $(0, t]$, the Poisson probability of observing k events is written as

$$\Pr(k;\lambda,t) = \frac{(\lambda t)^k e^{-\lambda t}}{k!} \tag{2.5}$$

For the time being, we delay giving a name to the parameter λ, or we could temporally call it the *rate of occurrence of events*.

The mean of the Poisson distribution in this case is

$$M = \lambda t \tag{2.6}$$

and its standard deviation is given by

$$\text{STD} = M^{1/2} = (\lambda t)^{1/2} \tag{2.7}$$

Note that the coefficient of variation (CoV) (which is the ratio of STD to the mean) of the Poisson distribution is

$$\text{CoV} = \frac{\text{STD}}{M} = \frac{1}{(\lambda t)^{1/2}} \tag{2.8}$$

and it gets smaller when the mean M increases. In engineering, the CoV is often referred to as the relative error.

In probability and statistics, as well as in reliability and risk analysis, the Poisson distribution is traditionally introduced as the limiting case of the binomial distribution, when the number of trials $n \to \infty$, but np is kept fixed.

It is clear that in order to keep np fixed, the probability p should decrease while n increases; therefore, the Poisson distribution is often referred to as the *distribution of rare events*, which can be confusing. We kindly ask our reader not to perceive the Poisson distribution as the *distribution of rare events*. We have to keep in mind that the Poisson distribution is the distribution of a *number* of something countable, such as the number of seeds in a watermelon, defects in a solid, or events in a given time interval. The term *event* belongs to the time domain (concretely defined time interval). Discussing the rare events, one has to specify a time unit, such as an hour (or a microsecond), as well as define a measure of the rareness (such as the time between events). Moreover, talking about rare events vs. frequent events, we cannot avoid taking into consideration our specific problem. For example, if one's car transmission fails three times during 4 years, no one will call these failures rare and the transmission reliable. On the other hand, if the owner of this car forgets to put on his or her badge three times during the same 4 years of her or his employment, these human failure events could be called rare ones. In the framework of risk analysis, choosing between rare and frequent might be a matter of the losses associated with the respective events. We will discuss this later, while introducing the Poisson *process*. So for the time being, please take the Poisson distribution as a model that turns out to be appropriate in numerous applications.

To illustrate that the Poisson probabilities are not small for quite realistic mean values, let's consider the Poisson distribution with mean $M = 3$. Table 2.1 and Figure 2.1 display the probability density function of this distribution.

First, we see that the probabilities of the numbers of events close to the mean of 3 are not small at all. Thus, the events, whose probabilities are

TABLE 2.1

Probability Density Function of Poisson Distribution with Mean $M = 3$

Number of Events, k	$Pr(k; M = 3)$
0	0.0498
1	0.1494
2	0.2240
3	0.2240
4	0.1680
5	0.1008
6	0.0504
7	0.0216
8	0.0081
9	0.0027
10	0.0008
20	$7.14 \cdot 10^{-11}$

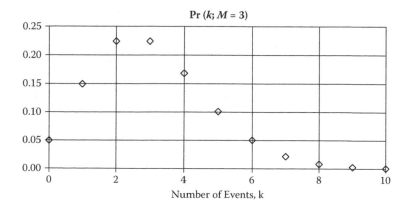

FIGURE 2.1
Probability density function of Poisson distribution with mean $M = 3$.

modeled by the distribution, are not rare. Is this example realistic or purely academic? It is realistic: e.g., it can be a distribution model of the number of failures of an air conditioner during a 10-year warranty coverage (not exactly $M = 3$, of course, but of the same order of magnitude). It also might be the distribution of the number of earthquakes in a given geographical region. Interested readers are referred to Rigdon and Basu (2000)—the only reliability book in which earthquake occurrence is successfully treated as failure occurrence in repairable systems.

It is difficult to name another discrete distribution having as many applications as the Poisson one does (from nuclear physics and genetics to earthquake studies and space engineering). For example, the NASA space shuttle BUMPER code for computing the damage from debris and from the impact of meteoroid particles uses the Poisson distribution as a model for these random events (Levin and Christiansen, 1997; Christiansen, Hyde, and Bernhard, 2004).

One of the earliest applications of the Poisson distribution to the problems, which now can be referred to as the risk analysis, is the long-living Bortkiewicz[*] data set of the number of horse-kick deaths of Prussian military personnel, which were collected for each of 14 corps in each of 20 years from 1875 through 1894 (Bortkiewicz, 1898; Preece, Ross, and Kirby, 1988). Note that the mean number of deaths per corps per year is about 0.7, which shows these fatal accidents among the Prussian Militärpersonen were not rare. For about 100 years, these data have been published and analyzed in many books and papers on probability and statistics.

We hope that we have convinced our reader not to think about the Poisson distribution as a distribution suitable for modeling only the rare events. However, we still have not suggested a simple physical model (similar to the tossing of a coin in the case of the binomial distribution) for the Poisson distribution.

Let's come back to our gambling reasoning. Now we need a die. A traditional die[†] is a cube. It has six sides (faces) marked on each side with the numbers 1, 2, ..., 6. There exist other types of dice with 4, 8, 12, and 20 faces. For our consideration, we have to make some assumptions. These assumptions will make our dice different compared to their real gambling counterparts.

First, we will consider our fair coin (the one we discussed at the beginning of this section) as a two-sided die, one side of which is marked by 0, and the other one by 1, and we are going to think about these two numbers as the random numbers of events (zero or one) occurring with equal

---

[*] Ladislaus von Bortkewitsch (1868–1931) was born in St. Petersburg (Russia) of Polish ancestry. After studies in Russia, he went to Germany. He received a PhD from the University of Gottingen, where he was a student of Wilhelm Lexis, to whom he later dedicated his monograph *Das Gesetz der kleinen Zahlen* (The Law of Small Numbers) (Bortkiewicz, 1898). In his publications, his name was variously spelled in the German form *Bortkewitsch* and the Polish form *Bortkiewicz* (Preece et al., 1988).

[†] The word *die* originates from Latin *datum* (with plural *data*), translated as "something given or played" (oxforddictionaries.com), which emphasizes the random nature of the data we deal with in our everyday life.

probabilities (1/2). Similarly, the cube die faces are now marked by 0, 1, ..., 5. Continuing our project, we come to the die with 20 faces, which is known as the *icosahedron*. Each of 20 equilateral triangle faces of the icosahedron we mark by 0, 1, 2, ..., 19.

So far, we have not made any unrealistic assumptions. In our first assumption (still, to an extent, realistic), we assume that we managed to make the icosahedron die with probabilities to land with a side facing upwards and marked by k (k = 0, 1, 2, ..., 19), given by the Poisson probability $\Pr(k; M)$ with a chosen value of mean M. For the sake of using the Poisson distribution illustrated by Figure 2.1 and Table 2.1, let's assume that M = 3. We hope that our reader is not confused by the integer value of this mean—generally speaking, the mean of a Poisson distribution can be any positive number. For the considered example, the respective probabilities of the outcomes 0, 1, 2, ..., 19 are given in Table 2.2.

Let us call our die the pseudo-Poisson icosahedron die. Why *pseudo*? Because the probabilities $\Pr(k; M = 3)$ do not add up to unity. The question is: How big is this error? Summing up all the probabilities in Table 3.2, one gets 0.99999999999, so we neglect any event with $k \geq 20$ with the probability, which is less than (1 – 0.99999999999)! Have you ever used the number π with so many decimals in your work?

The mean value of k for our pseudo-Poisson icosahedron die is given by

$$E(k;M) = \sum_{i=0}^{n-1} i \frac{M^i e^{-M}}{(i)!} \qquad (2.9)$$

In our example, the mean, calculated using the above formulae, is 2.99999999832, which is also close enough to the Poisson mean of 3.

TABLE 2.2

Icosahedron Die with the Poisson Probabilities to Land with a Side Facing Upwards and Marked by k

k	$\Pr(k; M = 3)$	k	$\Pr(k; M = 3)$
0	0.0498	10	$8.1015 \cdot 10^{-4}$
1	0.1494	11	$2.2095 \cdot 10^{-4}$
2	0.2240	12	$5.5238 \cdot 10^{-5}$
3	0.2240	13	$1.2747 \cdot 10^{-5}$
4	0.1680	14	$2.7315 \cdot 10^{-6}$
5	0.1008	15	$5.4631 \cdot 10^{-7}$
6	0.0504	16	$1.0243 \cdot 10^{-7}$
7	0.0216	17	$1.8076 \cdot 10^{-8}$
8	0.0081	18	$3.0127 \cdot 10^{-9}$
9	0.0027	19	$4.7569 \cdot 10^{-10}$

Looking at Table 2.1 and Figure 2.1, our attentive reader could notice that $\Pr(2; M = 3) = \Pr(3; M = 3)$. This is our small by-product result, revealing that if the Poisson mean M is an integer, the respective probability density function (PDF) values at M and $M - 1$ are equal, i.e.,

$$\Pr(M; M) = \Pr(M - 1; M)$$

$M = 1, 2, 3, \ldots$. We leave proving this property as a chapter exercise.

Thus, our pseudo-Poisson icosahedron die model looks like a rather precise one (approximation) for the Poisson distribution with $M = 3$. Now we can suggest the following imaginary physical model for the Poisson distribution with an arbitrary mean M. It is a die, having a large number of sides n (using the mathematical symbols $n \to \infty$). Each side of the die is marked by k ($k = 0$, $1, 2, \ldots$), and the probability that the die lands with a side facing upwards and marked by k is given by the Poisson probability $\Pr(k; M)$, i.e., (2.4).

Thinking about our die, one should keep in mind that it is not a fair die. For a fair die with n sides, the probability $p(k)$ is the same for each of n sides. This is a case of a perfect model for the so-called uniform distribution defined over a space of integers $0, 1, 2, \ldots, n - 1$, for which $\Pr(k; n) = 1/n$.

Now we are going to introduce the *homogeneous* Poisson process in the simplest possible way, needed to come to the exponential distribution. In Chapter 3, we will define this process in detail, in terms of the so-called stochastic point processes.

Let's assume we observe a *random point process* during a time interval $(0, t]$. The number k of events of the process, occurring in the interval $(0, t]$, is random, and it is distributed according to the Poisson distribution (2.5). It can be shown that in this case, the times between successive failures are distributed according to the exponential distribution defined in the next section. This random process is referred to as HPP. Strictly speaking, the HPP itself can be defined as a process, having the exponentially distributed (with the mean $M = \lambda t$) times between successive events.

2.1.2 Exponential Time-to-Failure Distribution

The first shock is fatal, life is exponential.

Now let us assume that simultaneously with the random process, we are observing a nonrepairable object. Every event of our process is a random shock, and each shock results in a failure of the object. Based on reliability tradition, we can refer to the nonrepairable object as the *component*. We also assume that at $t = 0$, the component is new.

Based on our model, the first shock causes the component failure, so that the component is functioning only when no shocks occur, i.e., the number of shocks $k = 0$. Thus, using the Poisson distribution (2.5), we can write the component reliability function as

$$R(t; \lambda) = \Pr(0; \lambda, t) = \exp(-\lambda t) \qquad (2.10)$$

On the component side, the process is terminated as soon as the first shock has occurred, so that, strictly speaking, the respective process should be called an *exponential* one, similar to the Weibull process discussed later in Chapter 3.

Using the expression (2.10), the cumulative distribution function (CDF) of the exponential distribution is written as

$$F(t; \lambda) = 1 - \exp(-\lambda t) \tag{2.11}$$

where $t \geq 0, \lambda > 0$.

Differentiating the above CDF, one gets the probability density function (PDF) of exponential distribution as

$$f(t; \lambda) = \lambda \exp(-\lambda t) \tag{2.12}$$

The most distinctive property of this distribution is related to its failure rate, which (based on the definition (1.4)) is constant, i.e.,

$$h(t) = \frac{f(t;\lambda)}{R(t;\lambda)} = \lambda \tag{2.13}$$

The multiplicative inverse of λ is the scale parameter of the exponential distribution. We will denote it by α. It can be shown (Barlow and Proschan, 1965, 1996) that the exponential distribution is the only time to failure (TTF) distribution having a constant failure rate.

Now we are coming to the notion of *aging*. The aging can be introduced in different ways. Let's recall the conditional reliability introduced in Chapter 1. Let's assume that the component TTF is exponentially distributed. Using the expression for the conditional reliability function (1.9) and the exponential reliability function (2.10), we find that for the object that survived to age t, the conditional reliability during the next time interval $(t, t + x]$ is

$$R(x|t) = \frac{R(t+x)}{R(t)} = \frac{e^{-\lambda(t+x)}}{e^{-\lambda(t)}} = e^{-\lambda x} \tag{2.14}$$

We see that the exponential conditional reliability (2.14) does not depend on age t. It means that an object having the exponentially distributed TTF does not age. This property of the exponential distribution is called the *memoryless* property (in the sense that the operating component has no memory of its age and past performance). This mathematical property does have an important reliability engineering implication—the objects, which are assumed to have exponentially distributed lives (TTF), do not require preventive maintenance, because they do not age.

The mean time to failure (MTTF) of the exponential distribution with reliability function (2.10) and its standard deviation (STD) are both equal to its scale parameter $\alpha = 1/\lambda$, such that the coefficient of variation (CoV) of the exponential distribution is

$$\mathrm{CoV} = \frac{\mathrm{STD}}{\mathrm{MTTF}} = 1 \qquad (2.15)$$

The coefficient of variation is a convenient and relative (and therefore *dimensionless*) measure of variability for positively defined random variables. In the following, we will need the coefficient of variation while discussing the aging and rejuvenating distributions.

Case Study

As an illustrative example of the exponential distribution, consider a population of 556,600 vehicles suffering 2,056 body panel paint scratches at various mileages observed throughout 20,000 miles. The scratches are caused by road stones pecking (against a body surface with nonrobust paint application), resulting in the degraded scratch resistance. Note that the estimated event probability at 20,000 miles is 0.00384 (or 0.384 of 1%). The vehicle manufacturer is interested in estimating the expected number of events over 60,000 miles in order to allocate a respective warranty budget.

Leaving aside the statistical aspects of constructing the estimate of the cumulative failure rate function (see Equation (1.6)), let us review it in Figure 2.2, where it is denoted by circles. The solid line through the circles is the cumulative failure rate function of the fitted exponential distribution. The correspondence between the circles and the line is an indication that the exponential distribution is *indeed* a reasonable probabilistic model for the underlying failure mileage distribution. This indication makes sense because the likelihood of a scratch depends neither on the accumulated mileage nor on the age of the vehicle, but rather on a random event of a road stone (of a certain mass) being thrown against the vehicle body (at a certain vehicle speed) such that the kinetic energy of the impact overcomes the paint resistance and leaves a scratch. Another evidence of the underlying exponential distribution is that the number of observed events per each 1,000-mile interval is approximately constant, irrespective of the ordinal number of the interval.

The slope of the fitted cumulative failure rate function in Figure 2.2 is an estimate of the failure rate function (2.13), which happens to be $1.946 \cdot 10^{-7}$ events/mile. Based on this, and according to (2.11), the event probability at 60,000 miles becomes 0.011, and the expected number of events over 60,000 miles is $0.011 \cdot 556,600 = 6122$.

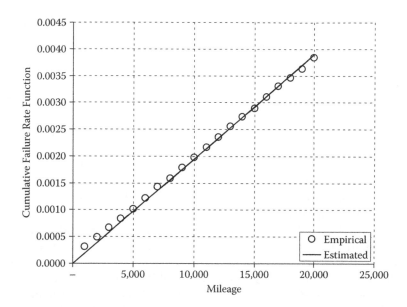

FIGURE 2.2
The empirical and theoretical (estimated) cumulative failure rate functions of the exponential distribution in the paint scratch example.

2.1.3 Gamma Time-to-Failure Distribution

If the kth shock is fatal, then life is gamma distributed with the index parameter k.

At this point, we are going to make a minor modification to the random shocks model, which resulted in the exponential time to failure (TTF) distribution in the previous section. Now, we assume that only the kth random shock is the lethal one for the nonrepairable object (component) of interest. In other words, the object survives $(k-1)$ successive random shocks, all arriving according to the homogeneous Poisson process (HPP), and it fails as soon as the kth shock occurs. The random TTF of our component T_k is the following sum of the k independent identically and exponentially distributed random variables t_i $(i = 1, 2, \ldots, k)$:

$$T_k = t_1 + t_2 + \ldots + t_k \tag{2.16}$$

The experiment behind evaluating T_k is as follows:

1. We set our chronometer at 0 time.

2. Simultaneously, we turn on our chronometer and start the HPP of the shock occurrences.

3. Simultaneously with the occurrence of the first shock (event), we record the time of its arrival t_1, stop the HPP, and set our chronometer at 0 time.

4. Turn on our chronometer, and start on the HPP of the shock occurrences.

Repeating steps 1 through 4 k times in our imaginary experiment, we get the random variable T_k.

We can also introduce our T_k in terms of reliability testing. Let's have a sample of k identical (from the reliability standpoint) components, which means that the time to failure of these components represents independent and identically distributed random variables. Moreover, we assume that the TTF distribution is the exponential one. Now, we put on a test of the first randomly chosen component and keep it testing until it fails, and the respective TTF is recorded as t_1. Then, we put on a test of the second randomly chosen component and run the same test to get the second TTF t_2. Continuing testing of the left $k - 2$ components in the same way, we will get our random variable T_k (2.16). One can suggest other reliability tests resulting in T_k. For example, we could test all the units simultaneously and wait until all of them have failed, recording each observed TTF. In the end, we have the same random variable T_k.

It can be shown that T_k is distributed according to the gamma distribution[*] with parameters k (nonrandom number of shocks resulting in failure) and λ ($1/\lambda$ is the mean time between successive shocks).

In the framework of the simple failure model, the component reliability function $R(t; k, \lambda)$ is equal to the probability that during time interval $[0, t]$ not more than $k - 1$ shocks occur; i.e.,

$$R(t;k,\lambda) = e^{-\lambda t} \sum_{i=0}^{k-1} \frac{(\lambda t)^i}{i!} \qquad (2.17)$$

The reliability function (2.17) is known as the reliability function of the gamma distribution.

The respective cumulative distribution function (CDF) of the gamma distribution obviously is

$$F(t;k,\lambda) = 1 - e^{-\lambda t} \sum_{i=0}^{k-1} \frac{(\lambda t)^i}{i!} \qquad (2.18)$$

---

[*] Strictly speaking, the distribution discussed in this section is called the Erlang distribution. The Erlang distribution is a particular case of the gamma distribution, the PDF of which is given by (2.19), in which the factorial is replaced by $\Gamma(k)$ and with arbitrary (not necessary integer) positive parameter k. More details are given below.

Taking the derivative of (2.18) with respect to time t, the gamma probability density function (PDF) can be obtained in the following form:

$$f(t;k,\lambda) = \lambda e^{-\lambda t} \frac{(\lambda t)^{k-1}}{(k-1)!} \qquad (2.19)$$

where $k = 1, 2, \ldots$. We leave the proof of (2.19) as a chapter exercise. It is not difficult to show (using the definition (1.4)) that the gamma distribution failure rate is given by

$$h(t;k,\lambda) = \frac{\dfrac{\lambda e^{-\lambda t}(\lambda t)^{k-1}}{(k-1)!}}{e^{-\lambda t}\displaystyle\sum_{i=0}^{k-1} \frac{(\lambda t)^i}{i!}} \qquad (2.20)$$

For the reliability model considered in this section, the gamma distribution parameter k is an integer taking on positive values 1, 2, ... In this case, the failure rate is nondecreasing; i.e., it is constant if $k = 1$ (exponential TTF), and it is increasing if $k > 1$.

It should be noted that, in the general case, the number of degrees of freedom k is not necessarily an integer. For arbitrary positive k, the PDF of the gamma distribution is written as

$$f(t;k,\lambda) = \lambda^k t^{k-1} \frac{e^{-\lambda t}}{\Gamma(k)} \qquad (2.21)$$

where $\Gamma(k)$ is the so-called *gamma* function, defined for all real positive k as the following integral:

$$\Gamma(k) = \int_0^\infty x^{k-1} e^{-x} dx \qquad (2.22)$$

and the failure rate is increasing if $k > 1$, and decreasing if $k < 1$.

The mean TTF (MTTF) of the gamma distribution with reliability function (2.17) and the TTF variance σ^2 are given by, respectively,

$$\text{MTTF} = \frac{k}{\lambda} \qquad (2.23)$$

$$\sigma^2 = \frac{k}{\lambda^2} \qquad (2.24)$$

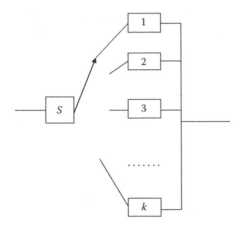

FIGURE 2.3
Standby system composed of k components and switch S.

The coefficient of variation of the gamma distribution is

$$\text{CoV} = \frac{\sigma}{\text{MTTF}} = \frac{1}{k^{1/2}} \tag{2.25}$$

The above-discussed random shocks model, in which we assume that the kth random shock is lethal, is not the only reliability model resulting in the gamma distributed TTF.

Let's consider a simple standby system composed of k identical (in the reliability meaning) components and a switch. The standby system is depicted in Figure 2.3.

The system begins functioning with component 1 in operational mode only—other $(k-1)$ components stay idle. At the moment when component 1 fails, switch S activates component 2, and so on, until the kth component fails. The time T_k, at which the kth component fails, is the system time to failure, which means that $T_k = t_1 + t_2 + \ldots + t_k$, where t_i $(i = 1, 2, \ldots, k)$ are the TTF of the system components. This system is called *one-out-of-k* standby system.

If we assume that these TTF t_i are independent and exponentially distributed with the same parameter λ, we will not make a mistake stating that the TTF of our standby system is distributed according to the gamma distribution with parameters k and λ, i.e., the system reliability function given by (2.17).

The main applications of the gamma distribution are related mostly to risk analysis problems such as modeling precipitation (including droughts), inventory control, and size of insurance claims. In Lawless (1982), we also find an example in which the gamma distribution is applied to a life data sample. The data are the survival times of 20 male rats that were exposed to a high level of radiation (p. 206).

FIGURE 2.4
The Ninth Wave (1850) by Ivan Aivazovsky from the Russian Museum in St. Petersburg, Russia. (From Wikipedia.)

Figure 2.4 shows a small photographic reproduction (from Wikipedia) of the gigantic painting (87 by 131 in.) by the famous Russian marine artist Ivan Aivazovsky. The painting is *The Ninth Wave* (1850) from the Russian Museum in St. Petersburg. It is one of the most well-known Russian paintings. It depicts a sea after a night storm and people in a desperate situation, trying to save themselves from a wrecked ship. The sky has warm tones, which are considered to symbolize a chance for the people to survive.

The painting title refers to the old nautical belief that the ninth wave is the most dangerous one (a kind of earliest maritime risk assessment). It should be mentioned that modern Fourier analysis of sea wave spectra does not confirm this old nautical belief. However, the ninth wave might serve as a simplified physical model of the gamma distribution with the parameter $k = 9$.

2.1.4 Normal (Gaussian) Time-to-Failure Distribution

Number of shocks before the fatal one increases, life (distribution) is getting normal

> Many years ago I called the Laplace–Gaussian curve the normal curve, which name, while it avoids an international question of priority, has the disadvantage of leading people to believe that all other distributions of frequency are in one sense or another "abnormal."
>
> **—Karl Pearson (1920)**

The normal distribution is, probably, the most well-known and commonly used distribution in applied and theoretical probability and statistics. However, as far as reliability is concerned, its popularity becomes limited. This might be explained in the following way. The most popular distributions in reliability, such as the exponential, Weibull, and lognormal ones, are the distributions of the positively defined random variables; i.e., they are used as the models for time to failure (or between failures), and time to failure is positive by common sense and by its definition. The normally distributed time to failure t is defined over the infinite interval ($-\infty < t < \infty$), so that there is always a nonzero probability that t is negative. The question is: How big is the probability of the negative time to failure when the normal distribution is applied to model real time-to-failure data? Another question is: For the normal distribution, could one suggest a simple probabilistic damage model, similar to the models we discussed in the previous sections, that resulted in the exponential and gamma distribution?

We are going to begin with the second question, but first we have to write the probability density function (PDF) and the cumulative distribution function (CDF) of the Gaussian distribution, for the sake of further references to this distribution. The PDF $f(t)$ and CDF $F(t)$ can be written as follows:

$$f(t; \mu, \sigma^2) = \frac{1}{\sqrt{2\pi}\sigma} \exp\left(-\frac{(t-\mu)^2}{2\sigma^2}\right) \tag{2.26}$$

$$F(t; \mu, \sigma^2) = \int_{-\infty}^{t} \frac{1}{\sqrt{2\pi}\sigma} \exp\left(-\frac{(\tau-\mu)^2}{2\sigma^2}\right) d\tau \tag{2.27}$$

where parameters μ, and σ^2 are, respectively, the mean and the variance of the normal distribution. The normal distribution has the traditional notation $N(\mu, \sigma^2)$. The PDF and CDF of the standard normal distribution (i.e., the normal distribution with $\mu = 0$ and $\sigma^2 = 1$) have the following standard notations:

$$\phi(t) = f(t; \mu = 0, \sigma^2 = 1) \tag{2.28}$$

$$\Phi(t) = F(t; \mu = 0, \sigma^2 = 1) \tag{2.29}$$

The normal distribution failure rate can be easily obtained using (2.26) and (2.27). The failure rate is an increasing function for any μ and σ^2, which makes the normal distribution a competitive model for aging objects.

In order to come to the normal distribution as the time to failure (TTF) distribution, we have to come back to the random shocks model, which resulted in the gamma TTF distribution in Section 2.1.3.

In the framework of that simple model, we considered a nonrepairable object (component) subject to random shocks, which occur according to the homogeneous Poisson process (HPP). We assumed that the kth arriving random shock is the lethal one for the object. The corresponding TTF distribution turned out to be the gamma distribution. Now we can ask the following question: If the number of shocks k needed to fail a component is getting larger and larger (i.e., $k \to \infty$), can we get an approximation for the component

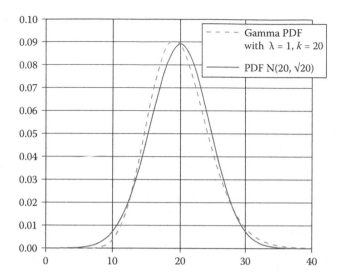

FIGURE 2.5
Probability density function of the gamma distribution with parameters $\lambda = 1$ and $k = 20$, and the PDF of its asymptotic counterpart—the normal PDF with a mean of 20 and a standard deviation of $\sqrt{20}$.

reliability function (2.17) and cumulative distribution function (2.18)? The answer is "yes we can." Based on the famous *central limit theorem* (Fischer, 2010; Hoyland and Rausand, 1994, 2004), one can state that the distribution of random sum (2.16), i.e., the distribution of $T_k = t_1 + t_2 + \ldots + t_k$, is getting close to the normal distribution with mean $\mu = k/\lambda$ and variance $\sigma^2 = k/\lambda^2$ (the reader recalls that $1/\lambda$ is the mean of the exponential distribution and $1/\lambda^2$ is its variance) as the number of shocks k increases ($k \to \infty$). Thus, if a given normal TTF distribution originates from the gamma distribution, its coefficient of variation $(\sigma/\mu) = 1/\sqrt{k}$ should be small enough.

It is interesting that in order to apply the central limit theorem we do not have to assume that the independent and identically distributed times between successive shocks t_i ($i = 1, 2, \ldots, k$) are *exponentially* distributed. We can simply assume that they are identically distributed and this distribution has a finite mean and a finite variance.

Figure 2.5 shows the PDF of the gamma distribution with parameters $\lambda = 1$ and $k = 20$ and the PDF of its asymptotic counterpart—the normal PDF with the mean of 20 and the standard deviation $\sqrt{20}$.

EXAMPLE 2.1
Nelson (1981, 1990) uses the normal distribution with mean $\mu = 6{,}250$ and $\sigma = 600$ years as a time-to-failure model of an insulation specimen. The respective coefficient of variation is $(\sigma/\mu) = 1/\sqrt{k} = 0.096$, based on which the number of shocks in the respective gamma distribution model can be evaluated as ≈ 108. The probability that the specimen's life is negative

is its cumulative distribution function at $t = 0$, i.e., $N(0; \mu, \sigma^2) \approx 1 \cdot 10^{-25}$, which is negligible.

2.1.5 Lognormal Time-to-Failure Distribution

Failure due to a growing fatigue crack.

There are at least two reasons behind the extensive use of this distribution in reliability and risk analysis. The first is the well-developed data analysis based on the normal origin of the lognormal distribution. The second reason is that the lognormal distribution results from the simple physical considerations in probabilistic fracture mechanics and the corresponding failure modes/processes (Mann, Schafer, and Singpurwalla, 1974; Provan, 1987).

Let $X_1 < X_2 < \ldots < X_n$ be a sequence of positive random variables that denote the size of a fatigue crack at successive stages of its growth. The so-called *proportional effect model* implies that during the *i*th stage, the crack size growth $X_i - X_{i-1}$ is *randomly* proportional to X_{i-1}, which is the size of the crack at the beginning of the stage; i.e.,

$$X_i - X_{i-1} = \pi_i X_{i-1} \qquad (2.30)$$

where the proportionality constant π_i is a random variable, and $i = 1, 2, \ldots, n$. The constants are assumed to be independently distributed, but not necessarily as having identical distributions for all n stages.

The above proportionality relation can be rewritten in the following form:

$$\frac{X_i - X_{i-1}}{X_{i-1}} = \pi_i \qquad (2.31)$$

In the limiting case, when $X_i - X_{i-1}$ tends to zero as n becomes large, using (2.31) we can write

$$\sum_{i=1}^{n} \pi_i = \sum_{i=1}^{n} \frac{X_i - X_{i-1}}{X_i} = \int_{X_0}^{X_n} \frac{dX}{X} = \ln(X_0) = \ln(X_0) \qquad (2.32)$$

or

$$\sum_{i=1}^{n} \pi_i = \ln\left(\frac{X_n}{X_0}\right) \qquad (2.33)$$

Since we assume that the proportionality constants π_i are independently distributed, it follows from the central limit theorem that their sum (2.33) converges to a normally distributed random variable. Thus,

$$\ln\left(\frac{X_n}{X_0}\right)$$

is asymptotically normally distributed, and

$$\frac{X_n}{X_0}$$

correspondingly has a lognormal distribution. In its turn, the relative size of the fatigue crack (with respect to the initial crack length)

$$\frac{X_n}{X_0}$$

is assumed to be proportional to the life length.

Now we are going to consider the traditional formal way of introducing the lognormal distribution. The lognormal distribution represents the distribution of a positive random variable whose logarithm follows the normal distribution. In other words, if x is normally distributed with a mean μ and a variance σ^2, then using the procedure for transformation of a random variable (see Appendix A), one can show that the positively defined random variable $t = e^x$ has a lognormal distribution with the following PDF:

$$f(t; \mu, \sigma) = \frac{1}{\sqrt{2\pi}\sigma t} \exp\left[-\frac{1}{2}\left(\frac{\ln(t)-\mu}{\sigma}\right)^2\right], \qquad t > 0 \qquad (2.34)$$

where μ and σ^2 are, respectively, the mean and the variance of $ln(t)$. Note that they are not the mean and the variance of the lognormal distribution. The mean and the variance of the lognormal distribution are given by

$$\mathrm{MTTF} = e^{\mu+\frac{\sigma^2}{2}} \qquad (2.35)$$

$$\mathrm{Var} = e^{2\mu+\sigma^2}\left(e^{\sigma^2}-1\right) = \left[\mathrm{MTTF}(t)\right]^2\left(e^{\sigma^2}-1\right) \qquad (2.36)$$

The cumulative distribution function (CDF) of the lognormal distribution is given by

$$F(t; \mu, \sigma) = \int_0^t \frac{1}{\sqrt{2\pi}\sigma\tau} \exp\left[-\frac{1}{2}\left(\frac{ln(\tau)-\mu}{\sigma}\right)^2\right] d\tau \qquad (2.37)$$

In terms of the standard normal distribution (2.29), the CDF (2.37) can be written as

$$F(t; \mu, \sigma) = \Phi\left(\frac{\ln(t) - \mu}{\sigma}\right) \tag{2.38}$$

and the respective reliability function as

$$R(t; \mu, \sigma) = 1 - \Phi\left(\frac{\ln(t) - \mu}{\sigma}\right) \equiv \Phi\left(\frac{\mu - \ln(t)}{\sigma}\right) \tag{2.39}$$

The important reliability measure—the median—has the following simple form for the lognormal distribution:

$$t_{0.5} = e^{\mu} \tag{2.40}$$

The lognormal distribution failure rate is given by

$$h(t; \mu, \sigma) = \frac{\dfrac{1}{\sqrt{2\pi}\sigma t} \exp\left[-\dfrac{1}{2}\left(\dfrac{\ln(t) - \mu}{\sigma}\right)^2\right]}{1 - \displaystyle\int_0^t \dfrac{1}{\sqrt{2\pi}\sigma\tau} \exp\left[-\dfrac{1}{2}\left(\dfrac{\ln(\tau) - \mu}{\sigma}\right)^2\right] d\tau}, \quad t > 0 \tag{2.41}$$

It has a specific shape (see Figure 2.6)—it is zero at the origin, then increases with time to a maximum, and after that it increases to zero (Nelson, 1981). This type of failure rate function is called *upside-down bathtub shaped*.

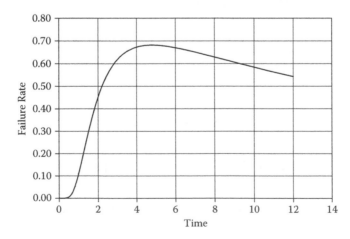

FIGURE 2.6
Failure rate of the lognormal distribution with $\mu = 1$ and $\sigma = 0.5$.

Besides the above-mentioned fatigue life problems, other applications of the lognormal distribution are related but not limited to modeling initial crack sizes (Laz and Hillberry, 1998), modeling uncertainties in the parameter λ of the exponential distribution and the respective HPP (Hoyland and Rausand, 1994, 2004), and reliability-based maintenance (e.g., modeling time to repair). Similar to the gamma distribution, the lognormal distribution is used as a distribution of claim size in the insurance business (Daykin, Penticainen, and Pesonen, 1995). The lognormal distribution of the number of loading cycles to failure is a typical assumption in fatigue life analyses of the so-called *S-N* or *Wöhler curves*, which are discussed later in this chapter. The distribution is also widely used in the probabilistic risk analysis to model the uncertainty of the failure rate λ of the exponential distribution.

Case Study: Fatigue Crack Modeling

As we discussed earlier, the lognormal distribution can arise from the simple physical considerations in probabilistic fracture mechanics. Figure 2.7 shows the empirical and fitted cumulative distribution functions of days in service needed for a crack to reach a given size (or failure criterion) on fatigue crack in a mechanical component. As it follows from the figure, the lognormal distribution ($\mu = 8.048$, $\sigma = 1.077$) provides a reasonable basis for modeling this failure mode.

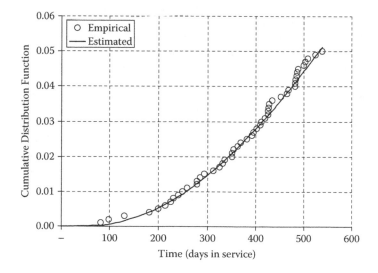

FIGURE 2.7
The empirical and fitted cumulative distribution functions of the lognormal distribution in the fatigue crack modeling example.

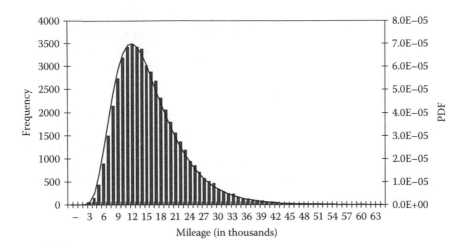

FIGURE 2.8
The histogram and the respective lognormal probability density function of the annual mileage of the owners of a mid-size sedan.

Case Study: Mileage Accumulation Modeling

A key element of estimating the cumulative distribution function from automotive warranty data (Krivtsov and Frankstein, 2004) is the probability of a vehicle exceeding a given mileage at a given time in service, which is a function of the vehicle's mileage accumulation. Many authors, including Lawless, Hu, and Cao (1995), Lu (1998), and Kaminskiy and Krivtsov (1999), indicate that the lognormal distribution is often an adequate choice for modeling mileage accumulation. Mileage accumulation can be modeled as a simple linear function $m = rt$, where m is cumulative mileage, t is time in service, and r is the mileage accumulation rate—a lognormally distributed random variable.

Figure 2.8 shows the histogram and the respective approximating lognormal probability density function of the annual mileage of the owners of a mid-size sedan. The estimated location and scale parameters of the lognormal distribution are 9.621 and 0.410, respectively. The good fit between the probability density function of the lognormal distribution and the empirical distribution (histogram) is an indication that the annual mileage may indeed be modeled as a lognormally distributed random variable. Using the expression for the lognormal mean (2.35), the average annual mileage for this population of vehicles is estimated to be 16,400 miles/year.

2.1.6 Weibull Time-to-Failure Distribution

As a generalization of the exponential distribution and/or as the weakest link model
Weibull distribution is possibly the most widely used distribution in modern reliability theory and life data analysis. In Chapter 3, we will introduce

the Weibull distribution through a shock model associated with a particular case of the nonhomogeneous Poisson process (the so-called *Weibull process*). At this point, we introduce the Weibull distribution as a simple formal generalization of the exponential distribution.

The reliability function of the exponential distribution (2.10) shows that the cumulative failure rate $H_{exp}(t)$ (recall expressions (1.6) and (1.7) from Chapter 1) of the distribution is given by the following simple linear function of time:

$$H_{exp}(t) = \lambda t \tag{2.42}$$

If we introduce a distribution with the cumulative failure rate given by the power function, we get the Weibull distribution in its standard form, from which the cumulative distribution function is given by

$$F(t; \alpha, \beta) = 1 - \exp\left(-\left(\frac{t}{\alpha}\right)^{\beta}\right) \tag{2.43}$$

where $t \geq 0$, parameter $\alpha > 0$ is the scale parameter, and $\beta > 0$ is the shape parameter. It is clear that the exponential distribution is a particular case of the Weibull distribution with the shape parameter $\beta = 1$.

The reliability function of the Weibull distribution obviously is

$$R(t; \alpha, \beta) = \exp\left(-\left(\frac{t}{\alpha}\right)^{\beta}\right) \tag{2.44}$$

For the time $t = \alpha$,

$$R(\alpha; \alpha, \beta) = \exp(-1),$$

and

$$F(\alpha; \alpha, \beta) = 1 - \exp(-1) \approx 1 - 0.3679 \approx 0.6321$$

By the definition of quantile, time $t = \alpha$ is 0.6321-level quantile (63.2th percentile) of the Weibull distribution. Note that this percentile/quantile does depend on the shape parameter β. This percentile interpretation of the scale parameter α explains why the parameter is often referred to as the *characteristic life* of the Weibull distribution.

The Weibull distribution probability density function is

$$f(t; \alpha, \beta) = \frac{\beta}{\alpha}\left(\frac{t}{\alpha}\right)^{\beta-1} \exp\left(-\left(\frac{t}{\alpha}\right)^{\beta}\right) \tag{2.45}$$

The failure rate of the distribution is given by the following expression:

$$h(t; \alpha, \beta) = \frac{\beta}{\alpha}\left(\frac{t}{\alpha}\right)^{\beta-1} \qquad (2.46)$$

The Weibull distribution failure rate is monotone, increasing if $\beta > 1$, decreasing if $\beta < 1$, and constant if $\beta = 1$ (the exponential distribution). If $\beta = 2$, the failure rate is the linear function of time, and in this case, the Weibull distribution is also known as the *Rayleigh distribution*.

Figures 2.9 and 2.10 illustrate the Weibull distribution probability density functions and failure rates.

The mean time to failure (MTTF) of the Weibull distribution and the TTF variance σ^2 are given by, respectively,

$$\text{MTTF} = \alpha\Gamma\left(1 + \frac{1}{\beta}\right) \qquad (2.47)$$

$$\sigma^2 = \alpha\left(\Gamma\left(1 + \frac{2}{\beta}\right) - \Gamma^2\left(1 + \frac{1}{\beta}\right)\right) \qquad (2.48)$$

where

$$\Gamma(a) = \int_0^\infty x^{a-1} e^{-x} dx, \qquad a > 0 \qquad (2.49)$$

is the gamma function, which was already introduced as (2.22).

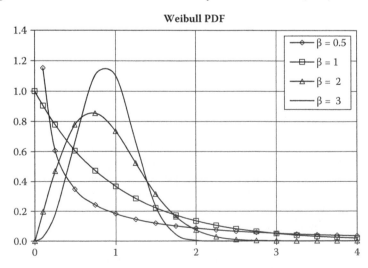

FIGURE 2.9
Weibull PDFs with scale parameter $\alpha = 1$ and different shape parameters β.

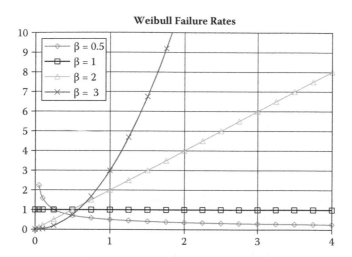

FIGURE 2.10
Weibull failure rate functions with scale parameter $\alpha = 1$ and different shape parameters β.

It is interesting that if the shape parameter β increases, the MTTF approaches the scale parameter (characteristic life) α, and the variance σ^2 approaches zero, which is illustrated by Table 2.3. Note also that the Weibull coefficient of variation (CoV) is greater than 1 for $\beta < 1$ (i.e., for a decreasing failure rate), it is 1 when $\beta = 1$, and it is greater than 1 if $\beta > 1$. The CoV also approaches zero as the shape parameter β increases.

TABLE 2.3

Mean, Variance, and Coefficient of Variation of Weibull Distribution

β	$\dfrac{\text{MTTF}}{\alpha}$	$\dfrac{\sigma^2}{\alpha^2}$	$\text{CoV} = \dfrac{\sigma}{\text{MTTF}}$
0.5	2	20	2.236
1	1	1	1.000
2	0.886	0.215	0.523
3	0.893	0.105	0.363
4	0.906	0.065	0.281
5	0.918	0.044	0.228
10	0.951	0.013	0.120
20	0.973	0.004	0.065
100	0.994	0.0002	0.014

2.1.6.1 Weakest Link Model

Let's consider a chain under a load. The chain consists of n links. All the links are assumed to have independent and identically distributed times to failure. Let the links have the time to failure (TTF) cumulative distribution function (CDF) $F(t)$, which we will refer to as the *parent distribution* while we discuss the weakest link model. The respective reliability function is $R(t) = 1 - F(t)$. Denote the links' TTF by $t_1, t_2, ..., t_i, ..., t_n$. It is clear that the TTF of our n-link chain T_n will be equal to the shortest of the TTF, i.e.,

$$T_n = \min(t_1, t_2, ..., t_n) \tag{2.50}$$

The random variable T_n is called the *smallest extreme value*. The reliability function of the n-link chain $R_{chain}(t)$ can be obtained as

$$R_{chain}(t) = \Pr(T_n > t) = \prod_{i=1}^{n} \Pr(t_i > t) = [R(t)]^n \tag{2.51}$$

and the respective CDF $F_{chain}(t)$ obviously is

$$F_{chain}(t) = 1 - [R(t)]^n \tag{2.52}$$

The above distribution is referred to as the *distribution of minima*.*

Now, we pose the following question: Can we find a good approximation to a distribution of minima in the following situations?

1. The number of links increases to infinity
2. The parent distribution $F(t)$ is unknown
3. The number of links is large but unknown

The answer to this question can be found in the so-called *theory of extreme value distributions*. Refer to the respective monographs on the theory and applications of extreme value distributions (Castillo, 1988; Kotz and Nadarajah, 2000). For our current discussion, all we need to know is that it has been mathematically proven by Fisher and Tippett (1928) and Gnedenko (1943) that there exist only three types of asymptotic distributions of maxima and three types of asymptotic distributions of minima. Type III of the asymptotic distributions of minima is the Weibull distribution.

The distributions introduced in the previous sections are the most popular in today's reliability and risk analyses. Many less popular lifetime distributions can be found in Lai and Min (2006).

---

* The reader familiar with the basic system reliability would notice that (2.51) is the reliability function of a *series system* composed of n identical components.

2.2 Classes of Aging/Rejuvenating Distributions and Their Properties

The failure rate that we introduced in Chapter 1 is closely related to the notion of aging. There exist different definitions of aging. The reliability definition discussed in this section has applications outside reliability theory and engineering; e.g., it turns out to be useful in describing and understanding the aging of biological systems, including human beings (Gavrilov and Gavrilova, 1991, 2006).

We call a population of identical* objects *aging* (*rejuvenating*) if its failure rate is, in a sense, increasing (decreasing). As we mentioned earlier, the only time-to-failure (TTF) distribution having a constant failure rate is the exponential one. Let's be more specific about aging. Many different *classes of time-to-failure distributions* are introduced based on properties of the failure rate, average failure rate, conditional reliability, and mean residual life, which we introduced in Chapter 1. The two most popular classes of aging distributions are the class of *increasing failure rate* (IFR) distributions and the class of *increasing failure rate average* (IFRA) distributions.

A TTF distribution is said to be IFR if its cumulative failure rate $H(t)$ (1.6) is *convex* (synonyms are *concave upwards, concave up*). Using the expression for reliability function (1.7) $R(t) = \exp[-H(t)]$, one can see that the statement *cumulative failure rate is convex* is equivalent to the statement that $-\ln(R(t))$ is convex, which is why the IFR distributions are also called *logarithmically convex distributions* (Barlow, Marshall, and Proschan, 1963).

The respective class of *rejuvenating distributions* consists of the *decreasing failure rate* (DFR) distributions. A TTF distribution is said to be DFR if its cumulative failure rate is concave.

Figure 2.11 illustrates the IFR (logarithmically convex) and DFR (logarithmically concave) distribution cumulative failure rates.

The notions introduced above are useful for the respective data analysis—if one needs to find out if a distribution is IFR or DFR, it is statistically easier to estimate $R(t)$ and then evaluate and plot $-\ln(R(t))$ than to estimate the respective failure rate $h(t)$.

Being a boundary distribution between IFR and DFR, the exponential distribution can be attributed to both classes (Barlow and Proschan, 1975; Leemis, 1995), in which case the IFR (DFR) distribution is defined as having a nondecreasing (nonincreasing) failure rate. One of the most counterintuitive results based on this definition of IFR distribution is related to a system of independent IFR components. It states that the *system of independent IFR*

---

* *Identical* from the reliability standpoint, i.e., population of objects having independent identically distributed failure times.

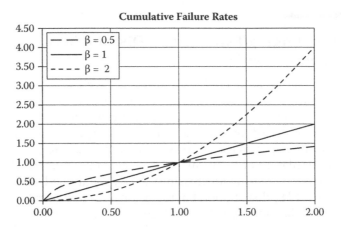

FIGURE 2.11
Cumulative failure rates for Weibull distribution with different shape parameters: $\beta = 2$ is an IFR distribution, $\beta = 0.5$ is a DFR distribution, and $\beta = 1$ is the exponential distribution (constant failure rate).

is not necessarily IFR distribution. The statement is usually illustrated by an example similar to the following (Leemis, 1995).

The so-called *parallel system* of two components is considered. By definition, a parallel system of n components functions if at least one of its components functions. Each of two components of our two-component system has an exponentially distributed time to failure, which means that the components belong to the IFR class. The reliability function of the system $R(t)$ is

$$R(t) = 1 - (1 - \exp(-\lambda_1 t))(1 - \exp(-\lambda_2 t))$$

It is suggested the reader derive the respective expressions for the system failure rate and average failure rate.

Let the first component have the exponential TTF distribution with $\lambda_1 = 1$ and the second component have the exponential TTF distribution with $\lambda_2 = 2$. The failure rate and average failure rate of the system are displayed in Figure 2.12.

The figure reveals that the failure rate of our system begins monotonically, decreasing after reaching a maximum at about $t = 1.5$ (please recall that the system reliability is low at that time). One could argue that this failure rate behavior might be related to the exponential time-to-failure distributions of the system components, and the exponential distribution, as we just said, is the boundary distribution between IFR and DFR and is attributed to both classes.

---

* Strictly speaking, the systems for which this is true are the so-called *coherent systems* defined in Appendix B. However, after reading this appendix, the reader will most probably agree that the adjective *coherent* is applicable to practically every real engineering system.

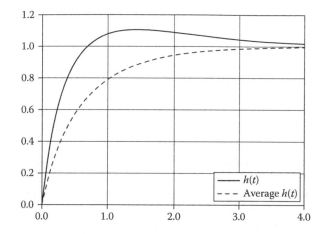

FIGURE 2.12
Failure rate $h(t)$ and average failure rate of the parallel system of two components having exponentially distributed lives with $\lambda_1 = 1$ and $\lambda_2 = 2$.

In order to reject this objection, let us consider a parallel system consisting of two components having strictly IFR time-to-failure distributions. Now, let the first component have the Weibull TTF distribution with the scale parameter $\alpha = 1$ (the same as in the previous example) and the shape parameter $\beta = 1.05$. The second component now has the Weibull TTF distribution with the scale parameter $\alpha = 0.5$ (as in the previous example) and the same shape parameter $\beta = 1.05$. The failure rate of the system is displayed in Figure 2.13. The figure reveals the same nonmonotonic failure rate of the system.

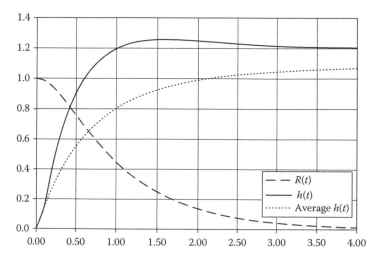

FIGURE 2.13
Failure rate $h(t)$, average failure rate, and reliability function $R(t)$ of a parallel system of two components having Weibull TTF distributions with increasing failure rates ($\beta = 1.05$).

Note that the average failure rate is monotonically increasing for the above-discussed systems, and so it makes sense to introduce the class of increasing failure rate average (IFRA) distributions as follows. A distribution is called IFRA if its average failure rate is nondecreasing in time (illustrated by Figures 2.12 and 2.13). Similarly, a distribution is called DFRA if its average failure rate is nonincreasing in time. It is clear that if a distribution is IFR (DFR), it is IFRA (DFRA) as well.

The Weibull distribution introduced in Section 2.1 is IFR (DFR) if its shape parameter $\beta \geq (\leq)$ 1. The gamma distribution is IFR (DFR) if its shape parameter $k > (<)$ 1. The normal distribution with any values of mean and variance is IFR distribution. As we remember, the lognormal distribution is a special one as far as its nonmonotonic failure rate is concerned. Therefore, the lognormal distribution is not an IFR distribution. It turns out that the lognormal distribution is neither IFRA nor DFRA (Barlow, 1979).

One of the most important results related to IFRA distributions states that *time to failure of any coherent system of independent IFRA components does have IFRA distribution* (Barlow and Proschan, 1975). Coming back to our example of the parallel two-component system, we can be sure that the system has the IFRA time-to-failure distribution, which is illustrated by Figures 2.12 and 2.13.

Every reliability resource discusses the so-called *bathtub* (*U-shaped*) failure rate time dependence. This shape of failure rate has been known and used in actuarial business[*] for more than 200 years. The interested reader is referred to James Dodson's lectures dated 1756 (Dodson, 1756). We will come back to the bathtub failure rate distributions in Section 2.3.6.

The classes of distributions introduced above are the most popular in today's reliability. A review of other classes of lifetime distributions (including the U-shaped one) can be found in Lai and Min (2006).

2.2.1 Damage Accumulation Models Resulting in Aging TTF Distributions

In Section 2.1 we considered some simple shock models resulting in the popular time-to-failure (TTF) distributions—exponential, gamma, normal, lognormal, and Weibull. The shocks (events) were occurring in time according to a point process. Each time, there was a fatal shock—the first one in case of the exponential distribution, the kth in the case of gamma distribution, etc. Now we are going to consider the models in a framework in which each random in-time shock is not necessarily fatal. Each shock will bring a random amount of damage to an object. These random damages from each shock are accumulated, and the object subjected to the shocks fails when the accumulated damage exceeds a certain threshold. We will see that the respective models result in the TTF distributions belonging to the classes of aging distributions introduced in the previous section.

---

[*] In actuarial applications, the failure rate is called the *rate of mortality*.

In order to discuss the damage accumulation models in more detail, we need to introduce the notion of *convolution*. In the given context, the distribution of the sum of two or more independent variables is the convolution of their individual distributions. Let us consider a simple case in which we have two positively defined independent random variables $x_1 > 0$ and $x_2 > 0$ with respective cumulative distribution functions (CDFs) $F_1(x_1)$ and $F_1(x_2)$. The probability distribution of the sum $y = x_1 + x_2$ is the convolution of their individual distributions. In terms of CDF, the convolution of y is given by

$$F(y) = \int_0^y F_1(y-x)dF_2(x) \equiv F_1 * F_2 \qquad (2.53)$$

If x_1 and x_2 have the respective probability density functions (PDFs) $f_1(x_1)$ and $f_1(x_2)$, the PDF of their sum $f(y)$ is given by the following convolution:

$$f(y) = \int_0^y f_1(y-x)f_2(x)dx \equiv f_1 * f_2 \qquad (2.54)$$

It is clear that the convolution of two independent random variables can be generalized to the case of $n > 2$ independent random variables. For example, the convolution of three random variables (the distribution of the sum of these three random variables) is the so-called *threefold convolution** of the convolution of two random variables (the distribution of their sum) and the third random variable.

2.2.1.1 Model I

The damage accumulation model is introduced as follows (Barlow and Proschan, 1975). An object is subject to shocks occurring in time according to the homogeneous Poisson process (HPP), which was introduced in Section 2.1. Each ith ($i = 1, 2, ...$) shock brings a random amount x_i of damage to the object. It is assumed that all x_i ($i = 1, 2, ...$) are independent, identically distributed random variables with the cumulative distribution function $F(x)$. The object fails when the accumulated damage exceeds a certain threshold X (see Figure 2.14).

Consider a time interval [0, t]. According to the HPP model, during this time interval, one can observe any number of shocks i ($i = 0, 1, 2, ..., \infty$). The probability that during the time interval [0, t] the object will be subjected to exactly i shocks is the Poisson probability of observing i events during this time interval; i.e., it is

---

* The convolution of two r.v. is called *twofold convolution* and the *onefold convolution* is the r.v. itself.

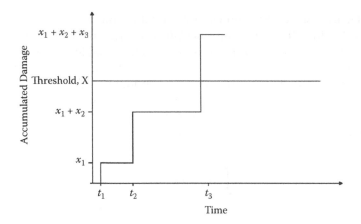

FIGURE 2.14
Random cumulative damage as function of shock arrival times. The object fails on arrival of third shock at t_3.

$$\Pr(i) = e^{-\lambda t}\frac{(\lambda t)^i}{i!}$$

The probability that after experiencing i shocks the object has the cumulative damage $(x_1 + x_2 + ... + x_i)$ not exceeding the threshold X (which means that our object survives after i shocks) is the i-fold convolution of the damage cumulative distribution function $F(X)$. Let's denote this convolution by $F^{(i)}(X)$. The probability that our object will not fail due to the accumulated damage obtained in time interval $[0, t]$, i.e., the object's reliability function, now can be written as

$$R\big(t; \lambda, F(X)\big) = \sum_{i=0}^{\infty} \Pr(i, \lambda t) F^{(i)}(X) = \sum_{i=0}^{\infty} e^{-\lambda t}\frac{(\lambda t)^i}{i!} F^{(i)}(X) \quad \text{for } 0 \le t < \infty \quad (2.55)$$

where $\Pr(i, \lambda t)$ is the probability of observing i shocks given by the Poisson distribution with mean λt. Note that for $i = 0$, the convolution $F^{(0)}(X)$ is defined as $F^{(0)}(X) = 1$.

For this cumulative damage model, Barlow and Proschan (1975) proved that for *any* damage distribution $F(x)$, the time-to-failure distribution (given by the reliability function (2.55)) is IFRA.

2.2.1.2 Model II

The above-considered model is expended to a bit more realistic one in which it is assumed that the shocks occur according to the same HPP process, but each successive shock becomes more effective in damaging the object.

However, the damages associated with each shock are still supposed to be statistically independent. The statement that each successive shock becomes more effective in damaging the object means that the CDF of the amount of damage caused by ith shock $F_i(x)$ is decreasing in i ($i = 1, 2, \ldots$) for each x. The probability that the object (having the fatal damage threshold X) survives after i shocks is given by the following convolution of the damage cumulative distribution functions: $F_0 X * F_1(X) * F_2(X) * \ldots * F_i(X)$, where $F_0(X) \equiv 1$.

The object's reliability function, as a function of time, now can be written as follows:

$$R\left(t; \lambda, F_1(X), F_2(X), \ldots\right) = \sum_{i=0}^{\infty} e^{-\lambda t} \frac{(\lambda t)^i}{i!} F_0(X) * F_1(X) * F_2(X) * \ldots * F_i(X)$$

(2.56)

$$\text{for } 0 \le t < \infty$$

According to the theorem proved by Barlow and Proschan (1975), the time-to-failure distribution (given by the reliability function (2.56)) is also IFRA.

The models introduced above have some important practical implications. If, based on prior knowledge or data analysis, we can accept the assumption that a distribution of interest is IFR or IFRA, we get some important tools for reliability estimation that are discussed in the following section.

2.2.2 Some Useful Inequalities for Reliability Measures and Characteristics for Aging/Rejuvenating Distributions

The following simple bounds and inequalities are given in terms of means, variances, and quantiles (percentiles) of IFRA (DFRA) distributions. It is worth recalling that any IFR (DFR) distribution belongs to the class of IFRA (DFRA) distributions.

2.2.2.1 Bounds Based on a Known Quantile (Barlow and Proschan, 1981)

Let t_p be pth quantile[*] (100pth percentile) of an IFRA (DFRA) time-to-failure distribution. Then the reliability function $R(t)$ satisfies the following inequality:

$$R(t) \begin{cases} \ge (\le) e^{-\alpha t} & \text{for } 0 \le t \le t_p \\ \le (\ge) e^{-\alpha t} & \text{for } t \ge t_p \end{cases}$$

(2.57)

where

$$\alpha = \frac{-\ln(1-p)}{t_p}$$

---

[*] By definition, if t_p is pth quantile, then $F(t_p) = 1 - R(t_p) = p$.

2.2.2.2 Bounds Based on a Known Mean (Barlow and Proschan, 1981)

Let μ be a mean of an IFR distribution. Then the reliability function $R(t)$ satisfies the following inequality:

$$R(t) \quad \begin{cases} \geq e^{-t/\mu} & \text{for } t < \mu \\ \geq 0 & \text{for } t \geq \mu \end{cases} \tag{2.58}$$

2.2.2.3 Inequality for Coefficient of Variation (Barlow and Proschan, 1981)

Let μ be a mean and σ^2 be the respective variance of an IFRA (DFRA) distribution. In this case, the coefficient of variation, i.e., σ/μ, satisfies the following inequality:

$$\sigma/\mu \leq (\geq) 1 \tag{2.59}$$

It is worth recalling that, for exponential distribution, the coefficient of variation is equal to 1. The inequality for the coefficient of variation is illustrated by the following example (Kaminskiy and Ushakov, 1995).

EXAMPLE 2.2

We are going to estimate the coefficient of variation for the samples of the well-known Birnbaum–Saunders data (Bogdanoff and Kozin, 1985). The 6061-T6 aluminum strips were cut parallel to the direction of the rolling of the sheet stock, mounted in simple supporting bearings, and deflected at the center with a Teflon clamp in reverse bending. The center was deflected 18 times per second and three stress amplitudes of 21, 26, and 31 ksi were used. The sample sizes were as follows: 101 specimens at 21 ksi, 102 specimens at 26 ksi, and 101 specimens at 31 ksi. All the specimens were tested to failure. The exposure (the analog of time to failure) is the number of cycles to failure. The sample means (in cycles), the sample variances (in cycles^2), and the sample coefficients of variation (dimensionless) are given in Table 2.4.

All of the sample coefficients of variation are less than 1, which indicates that these samples are from IFRA distributions.

TABLE 2.4

Means, Variances, and Coefficients of Variation for Birnbaum–Saunders Data

Stress Amplitude, ksi	Mean Number of Cycles to Failure/10^3	Variance of Number of Cycles to Failure/10^6	Coefficient of Variation
±21	1,400	152,881	0.28
±26	396	3,881	0.16
±31	134	502	0.17

2.2.3 Gini-Type Index for Aging/Rejuvenating Distributions

The index shows how different a given IFR (DFR) distribution is from the constant failure rate distribution (the exponential one).

As we mentioned earlier, the exponential distribution, which is the only distribution having a constant failure rate, plays a fundamental role in reliability. This distribution is the boundary one between the class of *increasing failure rate* (IFR) distributions and the class of *decreasing failure rate* (DFR) distributions. The distribution is closely related to the above-mentioned HPP. Indeed, in the framework of the HPP model, the distribution of the intervals between successive events observed during a time interval [0, t] is the exponential one with parameter λ equal to parameter λ of the respective Poisson distribution with mean λt.

In many practical situations, it is important to assess how far a given TTF distribution deviates from the exponential one, which can be considered a simple and therefore strong competing model. Note that if the exponential distribution turns out to be an adequate model, the respective object (component) is considered as nonaging, so it does not need preventive maintenance.

Some hypothesis testing procedures can be applied to help determine if the exponential distribution is an appropriate TTF model. In such situations, in principle, any goodness-of-fit test procedure can be used. Some of these tests for the null hypothesis (the times to failure are independent and identically exponentially distributed) appear to have good power against the IFR or DFR alternatives (Lawless, 2003).

Among such goodness-of-fit tests, one can mention the G-test, which is based on the so-called *Gini statistic* (Gail and Gastwirth, 1978). In turn, the Gini statistic originates from the so-called *Gini index* used in macroeconomics for comparing an income distribution of a given country with the uniform distribution (see Appendix C) covering the same income interval. The Gini index is used as a measure of income inequality (Sen, 1997). The index takes on values between 0 and 1. The closer the index value to zero, the closer the distribution of interest is to the uniform one. The interested reader could find the index values sorted by countries in List of Countries by Income Inequality (2011). It includes United Nations and CIA data.

Now we are going to introduce the so-called Gini-type index (Kaminskiy and Krivtsov, 2010a,b), helping to evaluate how far a given distribution deviates from the exponential one. Later, in Chapter 3, we will introduce a similar index for repairable systems modeled by point processes. In both cases, the index takes on values between –1 and 1. The closer the index's value is to zero, the closer the distribution of interest is to the exponential distribution. A positive (negative) value of the index indicates an IFR (DFR) failure time distribution. For the sake of simplicity, in the following, this Gini-type index will be referred to as *GT index* and denoted by *C*.

Consider a nonrepairable system (component) whose time to failure (TTF) distribution belongs to the class of the IFR distributions. Denote the failure rate associated with this distribution by $h(t)$. The respective cumulative hazard function (1.6) is then

$$H(t) = \int_0^t h(\tau)d\tau$$

and is concave upward.

Consider a time interval $[0, T]$. The cumulative failure rate function at T is $H(T)$, the respective TTF CDF is $F(T)$, and the reliability function is $R(T)$. Now, introduce h_{eff} as the failure rate of the exponential distribution with the CDF equal to the CDF of interest at the time $t = T$; i.e.,

$$h_{eff}(T) = -\frac{\ln(1-F(T))}{T}$$

In other words, the introduced exponential distribution with parameter h_{eff}, at $t = T$, has the same value of the cumulative hazard function as the IFR distribution of interest. See Figure 2.15.

The GT index, $C(T)$, is then introduced as

$$C(T) = 1 - \frac{\int_0^T H(t)dt}{0.5Th_{eff}(T)T} = 1 - \frac{2\int_0^T H(t)dt}{TH(T)} = 1 - \frac{2\int_0^T \ln(R(t))dt}{T\ln(R(T))} \qquad (2.60)$$

In terms of Figure 2.15, the index $C(T)$ is defined as 1 minus the ratio of areas A and $A + B$. It is easy to check that the above expression also holds

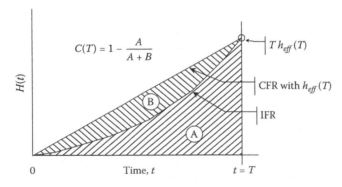

FIGURE 2.15
Graphical interpretation of the GT index for an IFR distribution.

for the decreasing failure rate (DFR) distributions, for which $H(t)$ is concave downward.

It is clear that the GT index $C(T)$ satisfies the following inequality: $-1 < C(T) < 1$. The index is positive for IFR distributions, negative for DFR distributions, and equal to zero for the constant failure rate distribution, i.e., the exponential distribution. Note that the suggested index is, in a sense, distribution-free; i.e., it is applicable to any continuous distribution used in reliability as a TTF distribution.

2.2.3.1 GT Index for the Weibull Distribution

For some TTF distributions, the GT index can be expressed in a closed form. For example, in the most important (in the reliability context) case of the Weibull distribution with the scale parameter α and the shape parameter β and the CDF of the form

$$F(t) = 1 - \exp\left(-\left(\frac{t}{\alpha}\right)^{\beta}\right)$$

the GT index can be found as

$$C = 1 - \frac{2}{\beta + 1} \tag{2.61}$$

It is worth noting that in this case, the GT index depends neither on the scale parameter α nor on the time interval T. Interestingly,

$$C(\beta) = -C\left(\frac{1}{\beta}\right),$$

which is illustrated by Table 2.5.

2.2.3.2 GT Index for the Gamma Distribution

Although not as popular as the Weibull distribution, the gamma distribution still has many important reliability applications. As was mentioned earlier, it is used to model a standby system consisting of k identical components with exponentially distributed times to failure; the gamma distribution is also the conjugate prior distribution in the Bayesian estimation of the exponential distribution.

TABLE 2.5

Gini-Type Index for Weibull Distribution as
Function of Shape Parameter β

Shape Parameter β	GT Index	TTF Distribution
5	0.6(6)	IFR
4	0.6	IFR
3	0.5	IFR
2	0.3(3)	IFR
1	0	CFR
0.5	−0.3(3)	DFR
0.3	−0.5	DFR
0.25	−0.6	DFR
0.2	−0.6(6)	DFR

Let's consider the gamma distribution with the CDF given by

$$F(t) = \frac{1}{\Gamma(k)} \int_0^{\lambda t} \tau^{k-1} e^{-\tau} d\tau = I(k, \lambda t) \tag{2.62}$$

where $k > 0$ is the shape parameter, $1/\lambda > 0$ is the scale parameter, and

$$I(k, x) = \int_0^x y^{k-1} e^{-y} dy$$

is the incomplete gamma function. Similar to the Weibull distribution, the gamma distribution has the IFR if the shape parameter $k > 1$, the DFR if $k < 1$, and the constant failure rate (CFR) if $k = 1$.

Using definition (2.60) and the CDF (2.62), the GT index for the gamma distribution can be written as

$$C(T) = 1 - \frac{2 \int_0^T \ln\left(1 - I(k, \lambda \tau)\right) d\tau}{T \ln\left(1 - I(k, \lambda T)\right)} \tag{2.63}$$

Table 2.6 displays $C(T)$ for the gamma distribution with $\lambda = 1$ evaluated at $T = 1$.

TABLE 2.6

GT Index for Gamma Distribution with $\lambda = 1$ and $T = 1$

Shape Parameter k	GT Index	TTF Distribution
5	0.623	IFR
4	0.543	IFR
3	0.428	IFR
2	0.258	IFR
1	0.000	CFR
0.5	−0.196	DFR
0.3	−0.285	DFR
0.25	−0.338	DFR
0.2	−0.375	DFR

2.3 Models with Explanatory Variables

The reliability models considered in the previous sections are expressed in terms of continuous or discrete distributions. They are a kind of physical model. For example, the most popular reliability distribution—the exponential—is also the nuclear decay model, in which the decay constant is, in reliability terms, the mean time to failure, and the *failure* is the decay event.

More complicated reliability models are needed in cases in which one is interested in reliability dependence on stress factors such as ambient temperature and humidity or voltage applied to a unit. This dependence is considered in the framework of the reliability models with *explanatory variables*, also known as *covariates*.

Such models are most often referred to as *accelerated life* (AL) *models*, a term that may be confusing because applications of these models are not necessarily limited to accelerated life testing. For the same product, a stress level can be considered elevated for one application, and it can be a normal stress level for another application. Nevertheless, we are going to use the term *accelerated life model*, as it is the most popular today.

2.3.1 Basic Notions

A reliability model (accelerated life [AL] reliability model) is defined as the relationship between the TTF distribution (or a parameter of the distribution) of an object and the stress factors applied to it, such as load, cycling rate, temperature, humidity, and voltage. Note that if no stress factors are considered, the model is simply reduced to the TTF distribution. In other words, any continuous TTF distribution discussed in the previous sections can be considered as a particular case of AL models.

Now we have to define *stress severity* in reliability terms. The stress severity in terms of reliability (or time-to-failure distribution) is expressed as follows. Let $R_1(t; z_1)$ and $R_2(t; z_2)$ be the reliability functions of the object under constant stress conditions z_1 and z_2, respectively. Note that stress condition z, in general, is a vector (set) of the stress factors (e.g., temperature and voltage). It is also important to keep in mind that a change in stress level results in the respective change of the TTF distribution, which is why we write $R_1(t; z_1)$ and $R_2(t; z_2)$, but if $z_1 = z_2$, then $R_1(t; z_1) \equiv R_2(t; z_2) = R(t; z_1) = R(t; z_2)$.

2.3.2 Stress Severity

The stress condition z_2 is called more severe than z_1 if for all values of time t the reliability of the object under stress condition z_2 is less than the reliability under stress condition z_1. That is,

$$R_2(t; z_2) < R_1(t; z_1) \tag{2.64}$$

2.3.3 Models with Constant Explanatory Variables (Stress Factors)

Time transformation function for the case of constant stress.

The time transformation function is easier to introduce in terms of respective TTF cumulative distribution functions (CDFs)—thus, we will use the CDFs $F_1(t; z_1) = 1 - R_1(t; z_1)$ and $F_2(t; z_2) = 1 - R_2(t; z_2)$.

For monotonic CDF $F_1(t; z_1)$ and $F_2(t; z_2)$, if constant stress condition z_1 is more severe than z_2, and t_1 and t_2 are the times at which $F_1(t_1; z_1) = F_2(t_2; z_2)$, there exists a function g (for all t_1 and t_2) such that $t_1 = g(t_2)$. Therefore,

$$F_2(t_2; z_2) = F_1[g(t_2), z_1] \tag{2.65}$$

because $F_1(t; z_1) < F_2(t; z_2)$, $g(t)$ must be an increasing function with $g(0) = 0$. The function $g(t)$ is called the *acceleration function* or the *time transformation function*. Figure 2.16 illustrates the concept of the time transformation function with $g(t) = \frac{1}{2} t$, which is the linear function; i.e., $F_2(t_2; z_2) = F_1[t_2/2, z_1]$. The last equality shows that time flies two times more quickly for the random variable (r.v.) distributed according to CDF F_2, compared to r. v. distributed according to CDF F_1.

Note that, in a general case, the time transformation function is a deterministic transformation of time to failure. Two main time transformations are considered in reliability theory. These transformations are known as the *accelerated life* (AL) *model* and the *proportional hazards* (PH) *model*. We begin with the AL model.

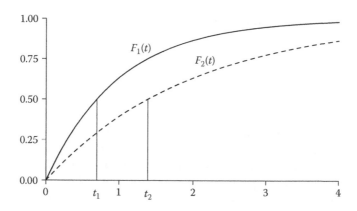

FIGURE 2.16
Time transformation function $g(t) = \frac{1}{2}t$.

2.3.3.1 Accelerated Life Model

Accelerated life model as time-to-failure scale transformation

In reliability engineering, the accelerated life model is the most popular type of reliability model with explanatory variables. For example, AT&T's Reliability Model (Klinger, Nakadar, and Menendez, 1990) is based on the AL model.

Without loss of generality, we assume the stress $z = 0$ for the normal (use) stress condition of an item of interest. Denote the time-to-failure CDF under the normal stress condition by $F_0(\cdot)$. By definition, in terms of the cumulative distribution functions $F(t; z)$ and $F_0(\cdot)$, the AL time transformation is given by the following relationship (Cox and Oakes, 1984):

$$F(t; z) = F_0[t\psi(z, A)] \tag{2.66}$$

where $\psi(z, A)$ is a positive function connecting time to failure with a vector of stress factors z, and A is a vector of unknown parameters estimated through the respective AL data analysis. For $z = 0$, $\psi(z = 0, A) = 1$; i.e., $F(t; z = 0) = F_0(t)$.

Relationship (2.66) is the scale transformation. It means that *a change in stress does not result in a change in the shape of the time-to-failure distribution function, but changes its scale only.* The scale transformation is widely used in physics—a Google search for "scale transformation in physics" would return millions of references.

Recalling the definition of the acceleration function, it is easy to see that relationship (2.66) can be written in terms of the acceleration function as follows:

$$g(t) = \psi(z, A)t \tag{2.67}$$

Relationship (2.66) is equivalent to the linear (with respect to time) acceleration function (2.67). The time-to-failure distribution of a device under the normal stress condition ($z = 0$) and the time-to-failure distribution of the identical device under a stress condition $z \neq 0$ are geometrically similar to each other. Such distributions are referred to as *belonging to the class of time-to-failure distribution functions, which is closed with respect to scale* (Leemis, 1995, 2009). The similarity property is widely used in physics and engineering. Because it is difficult to imagine that any change of failure modes or mechanisms would not result in a change in the shape of the failure time distribution, relationship (2.66) can also be considered a *principle of failure mechanism conservation or a similarity principle*, which states that if the failure modes and mechanisms remain the same over the stress domain of interest, then the respective time-to-failure distributions are similar to each other. The analyses of sets of real-life data often show that the similarity of time-to-failure distributions really exists, so violation of the similarity can indicate a change in failure mechanisms.

From the standpoint of random variable transformations, (2.67) is a simple linear transformation (like $y = bx$). Let t be a continuous random variable with mean $E(t)$ and variance $\sigma^2(t)$. Using the simple linear transformation, we can introduce accelerated time as a random variable $\tau = \psi(z, A)\, t$, where $\psi(z, A)$ is a constant. Our reader can easily check out (using, if needed, Appendix A, where the random variable transformations are discussed) that the mean $E(t)$ and variance $\sigma^2(t)$ of the new random variable τ are, respectively,

$$E(\tau) = \psi(z, A)\, E(t) \tag{2.68}$$

$$\sigma^2(\tau) = \psi^2(z, A)\, {}^2\sigma^2(t) \tag{2.69}$$

A useful property resulting from the above expressions is that the coefficient of variation (the ratio of the standard deviation to the mean) stays constant for any stress levels at which a given AL model is adequate:

$$\mathrm{CoV} = \frac{\sigma(\tau)}{E(\tau)} = const \tag{2.70}$$

Similarly, we can prove that the *variance of the logarithm of TTF stays constant at any stress condition in the domain where a given AL model holds*.

Let a random variable τ be the TTF at a stress z_1, and let t be the TTF at a stress z_2. Under the AL model (2.66), these random variables are related to each other as $\tau = \psi(z, A) t$ or, in a simplest form, as $\tau = kt$, where $k \equiv \psi(z, A)$ is a constant (i.e., it does not depend on time). Thus, we can write

$$\log(\tau) = \log(k) + \log(t)$$

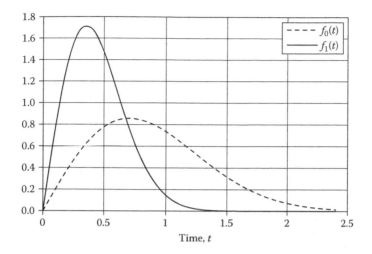

FIGURE 2.17
The PDF at the normal stress condition $f_0(t)$ is the Weibull PDF with the scale parameter equal to 1 and the shape parameter equal to 2. The PDF at the accelerated stress condition $f_1(t)$ was evaluated using (2.72) with $\psi = 2$, i.e., $f_1(t) = 2 f_0(2t)$.

Taking the variance of both sides of the above equality, one gets

$$\sigma^2[\log (\tau)] = \sigma^2[\log (t)] \tag{2.71}$$

Using the definition of AL model (2.66), the AL relationship for the TTF probability density functions (PDFs) can be obtained as

$$f(t; z) = \psi(z) f_0[t\psi(z, A)] \tag{2.72}$$

where $f_0(\cdot)$ is the time-to-failure PDF under the normal stress condition.

Figure 2.17 illustrates two probability density functions satisfying (2.72) with $\psi(z) = 2$ and $g(t) = 2t$. The PDF at the normal stress condition $f_0(t)$ is the Weibull PDF with the scale parameter equal to 1 and the shape parameter equal to 2. The PDF at the accelerated stress condition $f_1(t)$ was evaluated using (2.72) with $\psi = 2$, i.e., $f_1(t) = 2 f_0(2t)$. As a chapter exercise, we leave our reader to show that $f_1(t)$ is the Weibull PDF with the scale parameter equal to 0.5 and the shape parameter equal to 2.

The AL relationship for the $100p$th percentiles of time to failure, $t_p(z)$, can be obtained from (2.66) as

$$t_p(z) = \frac{t_p^0}{\psi(z, A)} \tag{2.73}$$

where $t_p{}^0$ is the $100p$th time to failure (TTF) percentile under the normal stress condition, i.e., for $z = 0$. A similar relation can be written for the corresponding random variables:

$$t(z) = \frac{t^0}{\psi(z, A)} \tag{2.73a}$$

The relationship (2.73) is the percentile AL reliability model (e.g., it can be an accelerated life model for the median), and it is usually written in the form

$$t_p(z, B) = \eta(z, B) \tag{2.74}$$

where B is a vector of the model parameters. The reliability models are discussed in Section 2.3.1.3, but for the time being, just take a look at the Arrhenius model (2.85), for which $z = T$ and the vector $B = (a, E_a)$.

The percentile AL reliability model is related to the relationship for percentiles (2.73) as

$$\eta(z, B) = \frac{t_p^0}{\psi(z, A)} \tag{2.75}$$

The corresponding relationship for the failure rates can also be obtained from (2.73a) as

$$\lambda(t; z) = \psi(z, A)\lambda_0[t\psi(z, A)] \tag{2.76}$$

where λ^0 is the failure rate under the normal stress condition, i.e., at $z = 0$.

Clearly, the relationship for percentiles (2.73) is the simplest one compared to the relationships for the cumulative distribution function (2.66), probability distribution function (2.72), and failure rate (2.76).

2.3.3.1.1 Cumulative Damage Models and Accelerated Life Model

Some of the cumulative damage models resulting in aging (say, IFRA) time-to-failure distributions, all of which were considered in Section 2.2.1, can be applied to the accelerated life (AL) model under quite reasonable restrictions. As an example, consider the Barlow and Proschan (1975) model, which we called Model I in Section 2.2.1.1.

In that section, we considered an object subjected to shocks occurring randomly in time. The shocks arrive according to the Poisson process with constant intensity (the homogeneous Poisson process). Each ith shock causes a random amount x_i ($i = 1, 2, \ldots, n$) of damage, where $x_1, x_2, \ldots, x_n$ are random variables distributed with a common CDF, $F(x)$, called a *damage distribution function*. The object fails when accumulated damage exceeds a threshold X. It has been shown by Barlow and Proschan (1975) that for *any* damage distribution function $F(x)$, the time-to-failure distribution function is IFRA.

Now, consider n identical objects under various stress conditions characterized by various Poisson shock intensities λ_i ($i = 1, 2, \ldots, n$) and by various damage distribution functions $F_i(x)$. In other words, we have n stress conditions, and each stress condition z_i is characterized by two factors: intensity λ_i and the damage distribution functions $F_i(x)$. It can be shown that the similarity of the corresponding time-to-failure distribution functions (the AL model) will hold for all these stress conditions, $z_i(\lambda_i, F_i(x))$, if they have the same damage distributions, i.e., if $F_i(x) = F(x)$. In a sense, we can say that the damage distributions $F_i(x)$ are associated with the object's failure modes, which, in the AL model factor domain, should be the same. A similar example from fracture mechanics is considered in Crowder et al. (1991).

2.3.3.2 Proportional Hazards Model

In the framework of the *proportional hazards* (PH) model or the *Cox model*, the basic relationship for the cumulative distribution function (CDF) analogous to (2.66) is given by

$$F(t;z) = 1 - \left[1 - F_0(t)\right]^{\psi(z,A)} \tag{2.77}$$

or in terms of the reliability function $R(t)$, it can be written as

$$R(t;z) = R_0(t)^{\psi(z,A)} \tag{2.78}$$

where $F_0(\cdot)$ and $R_0(\cdot)$ are, respectively, the time-to-failure CDF and the reliability function under the normal stress condition. *It should be kept in mind that the function $\psi(z, A)$ in the above-discussed accelerated life model and the function $\psi(z, A)$ discussed in this section on the PH model are numerically incomparable.* Generally speaking, the distribution transformation (2.77) is complex (Nelson, 1990), and for the PH model there is no simple relation for the $100p$th percentile of time to failure similar to the AL relationship (2.73).

The proper PH (Cox) model is known as the relationship for the failure rate (Cox and Oakes, 1984), which can be obtained from (2.77) and (2.78) as

$$\lambda(t;z) = \psi(z, A)\lambda_0(t) \tag{2.79}$$

where λ_0 is the failure rate under the normal stress condition, i.e., at $z = 0$, the function $\psi(z, A)$ is usually chosen as a log-linear function, i.e.,

$$\psi(z, A) \equiv \exp(a_1 z_1 + \ldots + a_n z_n) \equiv e^{a^\sim z}$$

and $z_1, \ldots, z_n$ are the explanatory variables (stress factors). This function results in a quite unusual time transformation; e.g., the reliability function (2.78) will take the form

$$R(t; z) = R_0(t)^{\exp(a_1 z_1 + \ldots + a_n z_n)} \tag{2.78a}$$

Comparing relationship (2.73) to (2.79), we come to an important conclusion about acceleration in the PH model—the acceleration function $\psi(z, A)$ acts multiplicatively on the failure rate, whereas in the AL model, it acts multiplicatively on the TTF percentile.

The PH model relationship for the TTF probability density functions (PDFs) can be obtained from (2.77) as

$$f(t; z) = \psi(z, A) f_0(t) R_0(t)^{\psi(z,A) -1} \tag{2.80}$$

Figure 2.18 illustrates the above relationship, which is analogous to Figure 2.17 for the AL model.

Note once more that generally speaking, the PH model time transformation is not a scale transformation of time, and it does not normally retain the shape of distribution. The function $\psi(z)$ no longer has a simple relationship to the acceleration function, nor does it have a clear physical meaning. That is why the PH model is not as popular in reliability applications as the AL model. However, the PH model is today the most widely used in biomedical life data analysis.

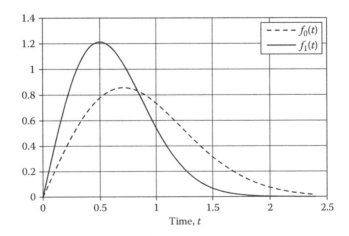

FIGURE 2.18
Proportional hazards model. The PDF at the normal stress condition $f_0(t)$ is the Weibull PDF with the scale parameter equal to 1 and the shape parameter equal to 2. The PDF at the stress condition $f_1(t)$ was evaluated using (2.80) with $\psi = 2$, i.e., $f_1(t) = 2 f_0(2t) R_0(t)$.

It should be mentioned that for the Weibull distribution (and only for the Weibull distribution) the PH model coincides with the AL model (Cox and Oakes, 1984).

2.3.3.3 Popular Accelerated Life Reliability Models

Without loss of generality, in this section we discuss the percentile AL models. Similar models can be written for MTTF or the failure rate. The most commonly used AL models for the percentiles (including median) of time-to-failure distributions are log-linear models. Two of such models are the power law model and the Arrhenius reaction model (Nelson, 1990).

The *power law percentile model* is given as

$$t_p(x) = \frac{a}{x^c}, \qquad a > 0, c > 0, x > 0 \qquad (2.81)$$

where x is a mechanical or electrical stress, c is a unitless constant, and a is a constant. The units of constant a are the product of the unit of time and the unit of x^c. In the reliability of electrical insulation and capacitors, x is usually the applied voltage.[*]

As far as mechanical applications of the power law model are concerned, it is widely used in the estimation of fatigue life as the analytical representation of the so-called *S-N* or *Wöhler curves*, where *S* is stress amplitude and *N* is life in cycles to failure, such that

$$N = kS^{-b} \qquad (2.82)$$

where b and k are material parameters estimated from test data. These curves are usually plotted in a log-log coordinate scale, which is illustrated by Figure 2.19.

Because of the probabilistic nature of fatigue life at any given stress level, one usually deals with not just one *S-N* curve, but with a family of *S-N* curves, so each curve is related to a probability of failure as the parameter of the model. These curves are called *S-N-P* curves, or curves of constant probability of failure on a stress vs. life plot. It should be noted that relationship (2.82) is an empirical model (Sobczyk and Spencer, 1992).

Another mechanical application of the power law model (2.82) is known as *Palmgren's equation*. The model is applied to the reliability of roller and ball bearings (Nelson, 1990). In this case, the percentile life t_p is measured in

---

[*] Generally speaking, relationship (2.81) is the power law function, which is widely used in physics. Simple examples are the Coulomb law and the gravitation law.

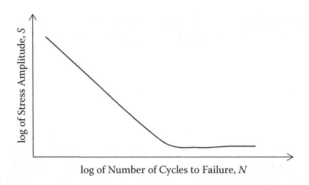

FIGURE 2.19
Typical *S-N* (Wöhler) curve.

millions of revolutions to failure, and x is the equivalent radial load, usually in pounds.

As far as electrical applications are concerned, the power law model is used to describe the so-called time-dependent breakdown of different dielectric materials. Most of the microelectronics publications on the time-dependent breakdown are related to the silicon dioxide thin films (e.g., see Wu and Suné, 2005).

It is interesting to note that the power law model has a property of *scale invariance*. Let us write (2.81) in a simple form as

$$f(x) = ax^b \tag{2.83}$$

Now let us *scale* the independent variable x by multiplying it by a constant factor s. We get

$$f(sx) = a(sx)^b = s^b f(x) \tag{2.84}$$

which shows that the scaling of the independent variable x results only in a proportionate scaling of the power law function itself.

The most popular reliability model explaining reliability dependence on temperature is the *Arrhenius model*,

$$t_p(T) = a\exp\left(\frac{E_a}{kT}\right) \tag{2.85}$$

where T is the absolute temperature (in Kelvin) under which the unit is functioning, E_a is the activation energy (usually in electron-volts, eV), and k is the Boltzmann constant. Originally introduced as a chemical reaction rate model, the Arrhenius model is now the most widely used model expressing the effect of temperature on reliability. The model forms the basis for a large portion of reliability (usually, failure rate) models of electronic components described in MIL-HDBK-217 and similar documents.

The following simple motivation for applying the Arrhenius chemical reaction rate model to reliability is from Wayne Nelson's monograph (1990). The Arrhenius *reaction rate* model is given by the next relationship between the rate of a simple (first-order) chemical reaction[*] and temperature:

$$r(T) = a' \exp\left(-\frac{E_a}{kT}\right) \tag{2.86}$$

If the failure of the object of interest is due to an accumulation of a critical amount of a chemical reaction product (as well as diffusion or a new phase growth), then this critical amount M_{cr} can be estimated as

$$M_{cr} = r(T)t \tag{2.87}$$

where t is a time to failure (life), so the time to failure is inversely proportional to the rate (2.86), which gives us a kind of justification of the reliability model (2.85).

Our next model is a combination of the power law model (2.81) and the Arrhenius model (2.85). The model has two explanatory variables—temperature and stress—and it is given by the following life relationship:

$$t_p(x, T) = a x^{-c} \exp\left(\frac{E_a}{T}\right) \tag{2.88}$$

where x is a mechanical or electrical stress. As a reliability model, it is used in the fracture mechanics of polymers as well as in the electromigration failures in the aluminum thin films of integrated circuits (IC). In the last case, the stress factor x is current density, and the power law model is referred to as *Black's equation* due to Black's pioneering work on IC reliability applications of the power law model (Black, 1969a,b). The model (2.88) is also used in the temperature-humidity tests (Nelson, 1990), in which case the stress x is the relative humidity, and the model is referred to as *Peck's relationship.*

Jurkov's model (Nelson, 1990) is another popular accelerated life reliability model. In the percentile form we use here, the model is given by the following relationship:

$$t_p(x, T) = t_0 \exp\left(\frac{E_a - \gamma x}{T}\right) \tag{2.89}$$

---

[*] In the case of first-order chemical reaction $A + B \rightarrow C$, the reaction rate equation is $r = [A][B]$. For the reaction $2D + E \rightarrow F$, the respective reaction rate equation is $r = [D]^2[E]$. The order of this reaction is 2 with respect to D, and with respect to E, the reaction order is 1.

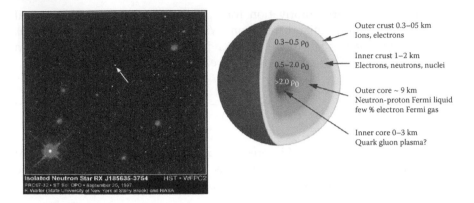

FIGURE 2.20
The first direct observation of a neutron star in visible light (the neutron star being en:RXJ185635-3754). (From F. Walter, Neutron Stars Are Really Small and Can't Be Seen by the Eye, State University of New York at Stony Brook, and NASA; ST Scl., http://en.wikipedia.org/wiki/File:Neutron_star_structure.jpg.)

with the same explanatory variables—temperature and stress—as in the previous model.

This model is considered (as any other model discussed in this section) an empirical relationship expressing the thermal fluctuation character of long-term strength, i.e., the durability under constant stress (Goldman, 1994; Regel, Slutsker, and Tomashevsky, 1974). For mechanical long-term strength, the parameter t_0 is a constant, which is numerically close to the period of thermal atomic oscillations (10^{-11} to 10^{-13} s); E_a is the effective activation energy, which is numerically close to the vaporization energy for metals and to chemical bond energies for polymers; and γ is a structural coefficient. The model is widely used for reliability prediction problems of mechanical and electrical (insulation, capacitors) long-term strength. It is also widely used in academic physics. For example, the model was recently used in the estimation of the breaking stress of neutron star crust (Chugunov and Horowitz, 2010). (See Figure 2.20.)

The a priori choice of a model (or competing models) is based on physical considerations. In its turn, statistical data analysis of accelerated life test results or collected field data, combined with failure mode and effects analysis (FMEA), can be used to check the adequacy of the chosen model or to determine the most appropriate model among the competing ones.

2.3.3.4 Popular Proportional Hazards Models

As mentioned in Section 2.3.3.2, the typical PH model with explanatory variable (2.79) is

$$\lambda(z, A) = \lambda_0(t)\exp(a_1 z_1 + \ldots + a_n z_n) \tag{2.79a}$$

where $z_1, \ldots, z_n$ are the explanatory variables and λ_0 is the failure rate under the normal stress condition, i.e., at $z = 0$. It should be noticed that in PH model applications, the researcher is often interested not in estimation of reliability itself, but in influence of the explanatory factors on reliability (Leemis, 1995). Application of the PH model is illustrated by the following examples.

EXAMPLE 2.3
Silicon Dioxide Breakdown (Part 1)

The proportional hazards (PH) reliability model was used in an attempt to reconcile voltage-to-breakdown data from a ramp-voltage (the time-dependent stress) test and time-to-breakdown data from a constant voltage (the constant stress) test (Chan, 1990). In this section, we consider the case of constant voltage stress only. The ramp-voltage (time-dependent stress) test is considered in Section 2.3.4.3.

The underlying time-to-failure distribution is assumed to be the Weibull one, i.e.,

$$R_0(t) = \exp\left\{\left[-\left(\frac{t}{a_0}\right)^\delta\right]\right\}$$

According to the PH relationship for reliability (2.78a),

$$R(t; z) = R_0(t)^{\psi(z, A)}$$

The function ψ is selected in the following simple form:

$$\psi(z, A) = e^{\beta z}$$

where the single stress factor (explanatory variable) z is $E_c - E_0$; E_c is constant electric field at test condition, E_0 is constant electric field at normal stress condition, and β is the only parameter.

Now, based on (2.78), the reliability function for a constant electric field at test condition E_c can be written as

$$R(t, z) = \exp\left\{\left[-\left(\frac{t}{\alpha_0}\right)\right]^\delta e^{\beta z}\right\}$$

$$R(t, z) = \exp\left\{\left[-\left(\frac{t}{\alpha_0}\right)\right]^\delta \left[e^{\frac{\beta z}{\delta}}\right]^\delta\right\}$$

$$R(t,z) = \exp\left(-\frac{t}{\alpha_c}\right)^{\delta}$$

where

$$\alpha_c = \alpha_0 \exp\left[-\left(\frac{\beta}{\delta}\right)(E_c - E_0)\right],$$

which is another Weibull distribution that could be anticipated in advance. The example for the case of linear increasing voltage (the time-dependent stress) from Chan (1990) is considered in Section 2.3.4.3.

2.3.4 Models with Time-Dependent Explanatory Variables (Stress Factors)

2.3.4.1 Accelerated Life Model with Time-Dependent Explanatory Variables

The models considered in the previous sections are applicable to constant explanatory variables (stress). The case of time-dependent explanatory variables is not only more general, but is also of more practical importance because its applications in reliability are not limited by accelerated life testing problems. As an example, consider the time-dependent stress analog of the power rule model (2.82).

The stress amplitude, S, experienced by a structural element usually is not constant during its service life; thus, the straightforward use of the S-N curve (Equation (2.82)) is not possible. In such situations the so-called *Palmgren–Miner rule* is widely used to estimate the fatigue life. The rule treats fatigue fracture as a result of a *linear accumulation* of partial fatigue damage fractions. According to the rule, the damage fraction Δ_i at any stress level S_i is linearly proportional to the ratio

$$\frac{n_i(S_i)}{N_i(S_i)},$$

where $n_i(S_i)$ is the number of cycles of operation under stress level S_i, and N_i is the total number of cycles to failure (life) under the *constant* stress level S_i. That is,

$$\Delta_i(S_i) = \frac{n_i(S_i)}{N_i(S_i)}, \qquad n_i \le N_i \tag{2.90}$$

The total accumulated damage, D, obtained under different stress levels $S_i(i = 1, 2, \ldots, n)$ is defined as

$$D = \sum_i \Delta_i(S_i) = \sum_i \frac{n_i(S_i)}{N_i(S_i)} \tag{2.91}$$

It is assumed that failure occurs if the total accumulated damage $D \geq 1$.

Accelerated life tests with time-dependent stress such as step-stress and ramp tests are also of practical importance. For example, one of the most common reliability tests of thin silicon dioxide films in metal-oxide semiconductor integrated circuits is the so-called ramp-voltage test. In this test, the oxide film is stressed to breakdown by a voltage that increases linearly with time (Chan, 1990).

However, it should be kept in mind that from a practical standpoint the accelerated life tests with time-dependent stress result in less accurate reliability predictions than their time-independent (constant) stress counterparts (Nelson, 1990). We also have to keep in mind that in order to get a physically meaningful model (with constant or time-dependent stress), the model should be related to one failure mode. The respective data analysis techniques helping to make sure that the failure mode is the same at different stress levels can be found in Gnedenko, Pavlov, and Ushakov (1999) and Modarres, Kaminskiy, and Krivtsov (2010).

Now we are going to consider a general case of the accelerated life (AL) model with time-dependent explanatory variables, so that in the end we will see that the Palmgren–Miner rule is a particular case of the AL model with time-dependent explanatory variables.

Let $z(t)$ be a time-dependent stress (i.e., vector of explanatory variables). We have to assume that the function $z(t)$ is integrable. In this case, the basic relationship for the cumulative density function (2.66) is replaced by the one suggested by Cox and Oakes (1984):

$$F\left[t;\, z(t)\right] = F_0\left[\Psi\left(t^{(z)}\right)\right] \tag{2.92}$$

where

$$\Psi\left(t^{(z)}\right) = \int_0^{t^{(z)}} \psi\left(z(\tau), A\right) d\tau\, ;\, t^{(z)}$$

is the time related to an item under the time-dependent stress condition $z(t)$, and A is a vector of AL model parameters.

Based on (2.92), the relationships for the time to failure (TTF) probability density function (PDF) and failure rate function can be obtained, respectively, as

$$f\left[t;z(t)\right] = \psi\left[z(t)\right] f_0\left[\psi\left(t^{(z)}\right)\right] \tag{2.93}$$

$$\lambda\left[t;z(t)\right] = \psi\left[z(t)\right] \lambda_0\left[\psi\left(t^{(z)}\right)\right] \tag{2.94}$$

The corresponding relationship for the $100p$th percentile of time to failure $t_p[z(t)]$ for the time-dependent stress, $z(t)$, can be obtained from (2.92) as

$$t_p^0 = \int_0^{t_p[z(t)]} \psi\big(z(\tau), A\big)\, d\tau \tag{2.95}$$

Using the relationship (2.74) for the AL percentile at constant stress, relationship (2.95) can be rewritten as

$$\int_0^{t_p[z(t)]} \frac{1}{t_p^0 \big\{\psi\big(z(\tau), A\big)\big\}^{-1}}\, d\tau = 1 \tag{2.96}$$

and, using (2.75), it can now be expressed in terms of the percentile reliability model for constant stress as

$$\int_0^{t_p[z(t)]} \frac{1}{\eta\big(z(\tau), B\big)}\, d\tau = 1 \tag{2.97}$$

where

$$\eta(z, B) = \frac{t_p^0}{\psi(z, A)}$$

is the AL percentile model for constant stress (2.75).

Note that relationship (2.97) is an exact nonparametric probabilistic continuous form of the Palmgren–Miner rule (2.91), which is discussed in detail in Section 2.3.4.2.

The integral equation (2.97) that we obtained has other important practical applications. Based on the percentile model for constant stress, it makes it possible to evaluate the percentile of time-to-failure distribution under a given integrable time-dependent stress $z(t)$. All we have to do is to replace the constant stress factors z in the model for constant stress with the respective time-dependent functions $z(t)$, and to solve the integral equation (2.97) with respect to $t_p(z)$. It can be successfully done if and only if the failure modes under the time-dependent stresses are the same as the failure modes under constant stresses.

EXAMPLE 2.4

Let's have an electronic device where the 100pth percentile life $t_p(U)$ dependence on constant applied voltage U is expressed through the power law model (2.81). Our problem is to evaluate the device's 100pth percentile life in the case when the applied voltage is linearly increasing in time τ; i.e., the applied voltage is $U(\tau) = k\tau$, where k is a constant. The $\eta(z,B)$ function from (2.97) can be written as

$$\eta(z, B) = \eta(U; a, c)) = \frac{a}{U^c} \tag{2.98}$$

where the only explanatory variable (stress) $z \equiv U$, and the vector of the model parameters $B = (a, c)$. In order to apply Equation (2.97) for time-dependent stress, we have only to replace the constant voltage U by the time-dependent $U(\tau) = k\tau$. Thus, using (2.97) we get

$$\frac{1}{a} \int_0^{t_p[U(t)]} (k\tau)^c d\tau = 1 \tag{2.99}$$

and we need to solve the above equation with respect to $t_p[U(t)]$, which is not difficult, because

$$\frac{1}{a} \int_0^{t_p[U(t)]} (k\tau)^c d\tau = \frac{k^c (t_p)^{c+1}}{a(c+1)} = 1 \tag{2.100}$$

so that our result is

$$t_p[U(t)] = \left[\frac{a(c+1)}{k^c} \right]^{\frac{1}{c+1}} \tag{2.101}$$

EXAMPLE 2.5

Instead of our previous unnamed electronic device, we are going to recall Nelson's transformer oil example (Nelson, 1990). All that we know about the transformer oil is that its life (in minutes) dependence on the applied *constant* voltage, U (in kV), is given by the power law model (2.81) as

$$\text{Life}(U) = \frac{1.2284 \; 10^{26}}{U^{16.3909}}$$

The normal (use) voltage is 15 kV, at which, according to the above equation, the life is $6.45 \; 10^6$ minutes, or about 12.3 years.

Let us assume that the transformer oil life is the median life, i.e., the 50th percentile of time to failure. Suppose that we want to estimate the same median life for the case when the applied voltage is linearly increasing in time τ; i.e., $U(\tau) = kt$, with the rate $k = 0.001$ kV per minute. Applying Equation (2.101), we find that the median life in the case of linearly increasing voltage is about $2.506 \; 10^4$ min or about 0.05 year.

This gives us the acceleration coefficient (the ratio of the life under the normal stress to the accelerated life) of 258.9. It is interesting to note that the voltage at the end of our accelerated test (simply speaking at the time when 50% of test units have failed) is $0.001 \times 2.506 \, 10^4 \approx 25$ kV, which is not very high compared to the normal stress voltage of 15 kV, and we might anticipate that the failure modes under normal and accelerated stress conditions are the same.

The power law model considered in this section is rather simple, which is why we could get the respective model with linearly increasing stress (which is also a simple time-dependent stress) in the closed form. In the general case, in order to get the model with time-dependent stress based on the respective model with constant stress, one has to apply numerical methods, which are illustrated by Example 2.6.

EXAMPLE 2.6
(Modarres et al., 2010)

The constant temperature Arrhenius reliability model for a component's 5th percentile of time to failure (2.85) is given by the following equation:

$$t_{0.05} = 2.590 \exp\left(\frac{0.400}{0.862 \times 10^{-4}(273 + T)}\right)$$

where $t_{0.05}$ is 5th percentile in hours, and T is temperature in °C. Find the 5th percentile of time to failure for the following thermal cycling, $T(t)$:

$T(t) = 25°C$ for $0 < t \leq 24$ h
$T(t) = 35°C$ for $24 < t \leq 48$ h
$T(t) = 25°C$ for $48 < t \leq 72$ h
$T(t) = 35°C$ for $72 < t \leq 96$ h

SOLUTION
An exact solution can be found as a solution for the following equation (based on relationship (2.97)):

$$\int_{0}^{t_p[T(s)]} \left[A\exp\left\{\frac{E_a}{b[T(s)+273]}\right\}\right]^{-1} ds = 1$$

Replacing the integral by the following sum, one gets

$$\sum_{i=1}^{k(t)} \delta_i + \delta(t^*) = 1$$

where $\delta_i = \delta$ is the damage accumulated in a complete cycle (48 h period); $\delta(t^*)$ is the damage accumulated during the last incomplete cycle, having duration t^*; k is the largest integer, for which $k\delta < 1$; and $\delta = \Delta_1 + \Delta_2$, where Δ_1 is the damage associated with the first 24 h of the cycle (under 25°C) and Δ_2 is the damage associated with the second part of the cycle (under 35°C). These damages can be calculated as

$$\Delta_1 = \frac{24}{A\exp\left\{\dfrac{E_a}{b[\,T_1+273\,]}\right\}}, \qquad \Delta_2 = \frac{24}{A\exp\left\{\dfrac{E_a}{b[\,T_2+273\,]}\right\}}$$

where $T_1 = 25°C$ and $T_2 = 35°C$. The numerical calculations result in $\Delta_1 = 1.6003 \times 10^{-6}$ and $\Delta_2 = 2.6533 \times 10^{-6}$. Thus,

$$\delta = \Delta_1 + \Delta_2 = 4.2532 \times 10^{-6}$$

The integer k is calculated as $k = [1/\delta] = 235100$, where $[x]$ is the greatest integer that does not exceed x. Estimate the damage accumulated during the last incomplete cycle, $\delta(t^*)$, as

$$\delta(t^*) = 1 - k\delta = 1 - 2.35 \times 10^{-4}(4.25 \times 10^{-6}) = 2.1510 \times 10^{-6} > 1.6003 \times 10^{-6}$$

which means that the last temperature in the profile is 35°C. Find t^* as a solution of the following equation:

$$\int_0^{t^*-24}\left[A\exp\left\{\frac{E_a}{b[35+273]}\right\}\right]^{-1}ds = 2.15 \times 10^{-6} - 1.60 \times 10^{-6}$$

which gives $t^* - 24 = 4.97$ (h). Finally, the exact solution is

$$t_p = 48\,k + 24 + 4.97 \approx 1.13 \times 10^7 \text{ (h)}$$

The correction obtained is negligible, but in the cases in which the cycle period is comparable with the anticipated life, the correction can be significant. Another important application of Equation (2.97) is estimation of parameters of AL models with constant stress based on AL testing with time-dependent stress, which is illustrated by Example 2.7.

EXAMPLE 2.7
(Modarres et al., 2010)

Let's have a ceramic capacitor, for which we would like to fit the constant stress AL model (2.88) for the 10th percentile in the following form:

TABLE 2.7

Ceramic Capacitor Test Results

Temperature, °K	Voltage U_0, V	Sample Time-to-Failure Percentile, h	Predicted Time-to-Failure Percentile, h
398	100	347.9	361.5
358	150	1,688.5	1,747.8
373	100	989.6	1,022.8
373	63	1,078.6	1,108.6

$$t_{0.1}(U,T) = aU^{-c} \exp\left(\frac{E_a}{T}\right)$$

where U is applied voltage and T is absolute temperature.

In order to fit the model, the following AL test plan using step-stress voltage in conjunction with constant temperature (as the accelerating stress factor) is used. Each test sample starts at a specified low voltage U_0 and is tested for a specified time Δt. Then the voltage is increased by ΔU, and the sample is tested at $U_0 + \Delta U$ during Δt. That is,

$$U(t) = U_0 + \Delta U \times En\left(\frac{t}{\Delta t}\right)$$

where $En(x)$ is the nearest integer not greater than x. The test will be terminated after the portion $p \geq 0.1$ of items fails. Thus, the test results are sample percentiles at each voltage–temperature combination. The test plan and simulated results with $\Delta U = 10$ V, $\Delta t = 24$ h are given in Table 2.7.

Our problem is to estimate the model parameters a, c, and E_a.

SOLUTION

For this example, we can write the following system of four integral equations (2.97) as follows:

$$\int_0^{(t_{0.1})_i} \frac{1}{a[U(\tau_i)]^{-c} \exp\left(\dfrac{E_a}{T_i}\right)} d\tau = 1, \quad i = 1, 2, 3, 4$$

or in a more explicit form as

$$\int_0^{347.9} \frac{1}{a[U(\tau)]^{-c} \exp\left(\dfrac{E_a}{398}\right)} d\tau = 1$$

where $U(\tau) = 100 + \Delta U En\left(\frac{\tau}{\Delta \tau}\right)$,

$$\int_0^{1688.5} \frac{1}{a[U(\tau)]^{-c} \exp\left(\dfrac{E_a}{358}\right)} d\tau = 1$$

where $U(\tau) = 150 + \Delta UEn\left(\frac{\tau}{\Delta\tau}\right)$,

$$\int_0^{989.6} \frac{1}{a[U(\tau)]^{-c} \exp\left(\dfrac{E_a}{373}\right)} d\tau = 1$$

where $U(\tau) = 100 + \Delta UEn\left(\frac{\tau}{\Delta\tau}\right)$, and

$$\int_0^{1078.6} \frac{1}{a[U(\tau)]^{-c} \exp\left(\dfrac{E_a}{373}\right)} d\tau = 1$$

where $U(\tau) = 63 + \Delta UEn\left(\frac{\tau}{\Delta\tau}\right)$.

Solving numerically, this system yields the following estimates for the model (2.88): $a = 2.23 \times 10^{-8}$ hV$^{1.88}$, $E_a = 1.32 \times 10^4$ °K, $c = 1.88$, which are close to the following values of the parameters used for simulating the data, $a = 2.43 \times 10^{-8}$ hV$^{1.87}$, $E_a = 1.32 \times 10^4$ °K, $c = 1.87$. The values of the percentiles predicted using the model are given in the last column of Table 2.7.

2.3.4.2 Accelerated Life Reliability Model for Time-Dependent Stress and Palmgren–Miner's Rule

It should be noted that our basic relationship (2.97)

$$\int_0^{t_p[z(t)]} \frac{1}{\eta(z(\tau), B)\}} = 1$$

connecting percentile life under constant stress with percentile life under time-dependent stress is an exact nonparametric probabilistic continuous form of the Palmgren–Miner rule (2.91) with $D = 1$. So, the problem of predicting life under time-dependent stress applying (2.97) is identical to the problem of predicting life through evaluating limiting cumulative mechanical damage addressed by the Palmgren–Miner rule. Moreover, there exists a useful analogy between mechanical damage accumulation and electrical breakdown. For example, we can recall that the power rule model and Jurkov's model are used as the life-stress relationships for both mechanical and electrical stresses.

In the theory of cumulative damage (Bolotin, 1989, 2010), a certain *damage measure D(t)* is introduced such that $0 \le D(t) \le 1$. It is assumed that $D(t)$ depends on its value at some initial time t_0 and on an external action $Q(t)$. Based on this assumption, the following general equation for $D(t)$ is postulated (Sobczyk and Spencer, 1992):

$$\frac{d[D(t)]}{dt} = f[D(t), Q(t)] \qquad (2.102)$$

where $f(D, Q)$ is a nonnegative function that satisfies the condition ensuring the existence and uniqueness of the solution of Equation (2.102). The equation is regarded as a *kinetic equation* for damage evaluation (Bolotin, 1989).

If the right side of Equation (2.102) does not depend on $D(t)$, the solution of the equation with the initial condition $D(0) = 0$ is the *linear damage accumulation model*, which is given by

$$D(t) = \int_0^t f[Q(\tau)]d\tau \qquad (2.103)$$

Similar to the Palmgren–Miner rule, the time T, at which the damage reaches its critical value, corresponds to the condition $D(T) = 1$. Using the notation $t(Q(t)) = 1/f(Q(t))$, the linear damage accumulation model at $t = T$ can be written as

$$\int_0^T \frac{d\tau}{t(Q(\tau))} = 1 \qquad (2.104)$$

Equation (2.104) formally coincides with our AL model with time-dependent stress (2.97), which means that the accelerated life model is the linear damage model. Being the linear damage accumulation model, it has an important reliability implication: it turns out that time is reversible under the AL model. It follows from the fact that the value of the integral in Equation (2.97) does not change when the stress function (history) $z(t)$ is changed to $z(t_p - t)$, where $t_p \ge t \ge 0$. The property of "reversible time" can be used to verify whether the AL assumptions are applicable to a given stress function. For example, a sample to be tested under a given time-dependent stress can be divided in two equal parts, so that the first subsample can be tested under the forward stress history, while the second subsample is tested under the backward stress.

2.3.4.3 Proportional Hazards Model with Time-Dependent Explanatory Variables

The basic proportional hazards (PH) model in the case of time-dependent explanatory variables (stresses) is given as the following relationship for the failure rate:

$$\lambda(t; z(t)) = \psi(z(t), A)\lambda_0(t) \tag{2.105}$$

where $z(t)$ is, in general, a vector (set) of stress conditions and at least some of them are time dependent, and $\lambda_0(t)$ is the failure rate under the normal (nominal) stress condition, i.e., at $z = 0$ and $\psi(0) = 1$. We can see that the case of the constant stress (2.79) is a particular case of (2.105).

In the following, the most popular case of log-linear $\psi(z(t), A)$ function is considered, i.e.,

$$\psi(z, A) \equiv \exp\left(a_1 z_1(t) + \cdots + a_n z_n(t)\right) \equiv e^{a^{\cdot}z(t)} \tag{2.106}$$

so that (2.105) can be written as

$$\lambda(t; z(t)) = \lambda_0(t) e^{a^{\cdot}z(t)} \tag{2.107}$$

where $z(t) \equiv z_1(t), \ldots, z_n(t)$ are the time-dependent explanatory variables.

Recalling relationship (1.5), the respective relationship for the reliability function can be written as

$$R(t; z(t)) = \exp\left(-\int_0^t \lambda_0(\tau) e^{a^{\cdot}z(\tau)} d\tau\right) \tag{2.108}$$

EXAMPLE 2.8
Silicon Dioxide Breakdown (Part 2)

We are continuing to consider the example on the silicon dioxide breakdown (Chan, 1990) from Section 2.3.3.4. Now we are constructing a model for the ramp-voltage test, under which the oxide film is stressed to breakdown by a voltage that increases linearly with time. Now our random variable of interest is not the time to breakdown, as it was in Section 2.3.3.4, but the *voltage to breakdown*.

The time-dependent explanatory variable $z(t)$ is a linear function of time

$$z(t) = E_s(t) - E_0 = kt - E_0$$

where E_0 is the constant electric field at a normal stress condition and $E_s(t) = kt$ is linearly increasing with time t ramp voltage.

The same underlying Weibull distribution as in the first part of this example is assumed, i.e.,

$$R_0(t) = \exp\left\{\left[-\left(\frac{t}{\alpha_0}\right)^{\delta}\right]\right\}$$

which corresponds to the following failure rate function:

$$\lambda_0(t) = \frac{\delta}{\alpha_0}\left(\frac{t}{\alpha_0}\right)^{\delta-1}$$

Applying (2.109), we get

$$\ln\left[R(t;z(t))\right] = -\int_0^t \lambda_0(\tau)e^{\beta z(\tau)}d\tau$$

$$= -\frac{\delta}{\alpha_0}\int_0^t \left(\frac{\tau}{\alpha_0}\right)^{\delta-1}e^{\beta z(\tau)}d\tau$$

Changing the integration variable $\tau = E_s/R$, it can be written as

$$\ln\left[R(E_B;k)\right] = \frac{-\delta}{(\alpha_0 k)^{\delta}}e^{-\beta E_0}\int_0^{E_B} E_S^{\delta-1}e^{\beta E_S}dE_S$$

where $E_B = kt$ is called the *breakdown field*.

The integral in the above relationship, in general, cannot be expressed in a closed form, and the resulting reliability function

$$R(E_B) \equiv R(E_B;k) = \exp\left(\frac{-\delta}{(\alpha_0 k)^{\delta}}e^{-\beta E_0}\int_0^{E_B} E_S^{\delta-1}e^{\beta E_S}dE_S\right)$$

does not look simple and easily interpretable.

2.3.5 Competing Failure Modes and Series System Model

The models considered in this section, traditionally, are not referred to as the models with explanatory variables. However, sometimes they provide much more insights about the failure behavior than the simple time-to-failure distributions discussed in Section 2.1.

Consider a *series system* depicted in Figure 2.21. The system consists of k components, and it is functioning if and only if all of its n components are functioning. Similarly, we can consider an object having k *competing failure*

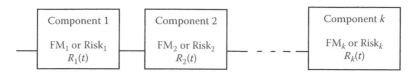

FIGURE 2.21

A series system composed of k components, or an object having k competing FMs.

TABLE 2.8

Series System/Competing Risks (FM) Model Assumptions

Series System Model	Competing Risks (FM) Model
The system fails when the first component failure occurs.	The object fails when the first failure due to any or all the competing risks (FM) occurs.
Time to failure of each component (FM) is independent of TTF of any other component (FM), so that the system TTF distribution due to all of its components (FM) is given by any of the equations (2.109)–(2.112).	Time to failure due to a given risk (FM) is independent of TTF due to every other risk (FM), so that TTF distribution due to all the failure modes is given by any of the equations (2.109)–(2.112).

modes (mechanisms) or *competing risks*. The object fails when the first failure of these competing failure mode (FM) occurs.

The model we are going to discuss can have two names: the *series system model* (in the context of system reliability engineering) and the *competing risks (failure modes) model* (in the physics of failures and biomedical applications). These mathematically identical models are based on the same assumptions, which are formulated in Table 2.8.

The reliability function, TTF cumulative distribution function (CDF), TTF probability density function (PDF), and failure rate for the competing risks model are given, respectively, by the following formulas:

$$R_{\text{comp}}(t) = \prod_{i=1}^{k} R_i(t) \tag{2.109}$$

$$F_{\text{comp}}(t) = 1 - \prod_{i=1}^{k} \left\{ 1 - F_i(t) \right\} \tag{2.110}$$

$$f_{\text{compt}}(t) = \sum_{j=1}^{k} f_j(t) \prod_{\substack{i=1 \\ i \neq j}}^{k} \left\{ 1 - F_i(t) \right\} \tag{2.111}$$

$$\lambda_{\text{comp}}(t) = \sum_{i=1}^{k} \lambda_i(t) \tag{2.112}$$

where $R_i(t)$, $F_i(t)$, $f_i(t)$, and $\lambda_i(t)$ are, accordingly, the reliability function, TTF CDF, TTF PDF, and failure rate related to the ith risk (FM_i) or ith component of a series system, and $i = 1, 2, ..., k$.

2.3.6 Competing Risks Model vs. Mixture Distribution Model

A rival model for the competing risks model might be the so-called *mixture distribution model*, which is the mixture of k distributions, each of which is associated with a respective FM. If $F_1(t)$, $F_2(t)$, ..., $F_k(t)$ are the cumulative distribution functions (CDFs) and the positive constants $p_1, p_2, ..., p_k$ are such that

$$\sum_{i=1}^{k} p_i = 1$$

then the mixture distribution CDF is

$$F_{mix}(t) = \sum_{i=1}^{k} p_i F_i(t) \tag{2.113}$$

The reliability function is given by

$$R_{mix}(t) = \sum_{i=1}^{k} p_i R_i(t) \tag{2.114}$$

The positive constants $p_1, p_2, ..., p_k$ are referred to as the *weights* or *mixture weights*.
Differentiating (2.113), we obtain the following expression for the mixture PDF:

$$f_{mix}(t) = \sum_{i=1}^{k} p_i f_i(t) \tag{2.115}$$

where

$$f_i(t) = \frac{dF_i(t)}{dt}$$

is the probability density function (PDF), which is assumed to exist. It should be reiterated, that in the given context, we think about the k distributions as the TTF distributions related to the k failure modes (FMs).

Unlike the competing risks failure rate, the mixture failure rate is not expressed in a simple form. It is given by

$$\lambda_{mix}(t) = \frac{\sum_{i=1}^{k} p_i f_i(t)}{\sum_{i=1}^{k} p_i R_i(t)} \qquad (2.116)$$

In the following, we are going to discuss in more detail the mixture distribution model, consisting of two distributions. In this case, Equations (2.113) and (2.115) are traditionally written, respectively, as

$$F_{mix}(t) = pF_1(t) + (1-p)F_2(t) \qquad (2.117)$$

$$f_{mix}(t) = pf_1(t) + (1-p)f_2(t) \qquad (2.118)$$

The mixture PDF (2.118) is illustrated by Figure 2.22, in which the mixture weight $p = 0.2$; the first distribution is the Weibull with the scale parameter α equal to 1 and the shape parameter is equal to 2. The second distribution is also the Weibull distribution with the scale parameter α equal to 5 and the shape parameter equal to 4, so both distributions are aging.

The respective mixture cumulative distribution function (2.117) is illustrated by Figure 2.23.

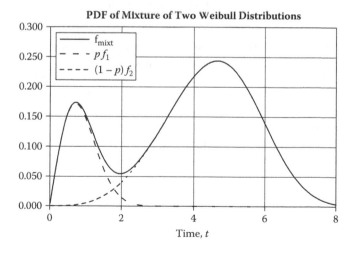

PDF of Mixture of Two Weibull Distributions

FIGURE 2.22
Probability density function of mixture of two distributions. The first distribution is a Weibull distribution with scale parameter α equal to 1 and shape parameter of 2. The second distribution is also a Weibull one with scale parameter α equal to 5 and shape parameter of 4. The mixture weight $p = 0.2$.

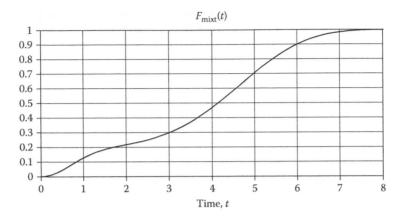

FIGURE 2.23

Cumulative distribution function of mixture of two distributions. The first distribution is a Weibull distribution with scale parameter α equal to 1 and shape parameter of 2. The second distribution is also a Weibull one with scale parameter α equal to 5 and shape parameter of 4. The mixture weight $p = 0.2$.

A typical physics of failure mixture model analysis includes using the so-called Weibull probability paper. The Weibull probability paper is based on the following property of the Weibull distribution (see Section 2.1.6). Taking logarithm twice from the Weibull CDF, one gets

$$\ln\left[-\ln\left(F(t)\right)\right] = \beta \ln(t) - \beta \ln(\alpha)$$

which is a straight line in the respective coordinate system. The Weibull probability paper for the mixture distribution considered above is shown in Figure 2.24. The picture reveals a presence of the two failure modes with different time-to-failure distributions, with the second distribution having a greater scale and shape parameters.

Now let's note that $pF_1(t)$ in (2.117) is the probability that the FM_1 failure occurs in time interval $(0, t]$, and that $(1 - p)F_2(t)$ is the probability that the FM_2 failure occurs in the same time interval. Note that

$$\lim_{t \to \infty} pF_1(t) = p$$

and

$$\lim_{t \to \infty} (1 - p)F_2(t) = 1 - p$$

For the mixture model, it might be possible to approximately specify a time interval during which a given FM is a dominating one. For example,

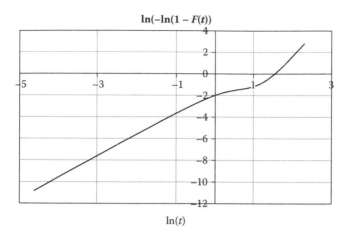

FIGURE 2.24
Weibull probability paper for mixture of two distributions, of which the probability density function and cumulative distribution function are depicted in Figures 2.22 and 2.23, respectively.

if μ_1 is the mean time to failure of FM_1 and it is somewhere inside the time interval $(0, t_0]$ and the mean time to failure of FM_2 $\mu_2 \gg t_0$, we can state that FM_1 is a dominating failure mode in $(0, t_0]$. For example, for the mixture of two Weibull distributions considered above with $p = 0.2$, using the expression for Weibull mean (2.47), we conclude that the mean time to failure of FM_1 is 0.89 and the mean time to failure of FM_2 is 4.53. Let's choose $t_0 = 1$. In this case, the probability that an FM_1 failure occurs in time interval $(0, 1]$ is $pF_1 \approx 0.126$, while the probability that an FM_2 failure occurs during the same time interval is only $(1 - p)F_2(1) \approx 0.00128$, which shows that FM_1 is the dominating failure mode during the time interval $(0, 1]$. This dominance is also revealed by Figure 2.22. Continuing in a similar way, we can show that in the time interval $(1, 8]$, the second failure mode is dominant.

In the case of the competing risks model, the dominating failure mode is the one that has the shorter failure time, which is illustrated by Figure 2.25. Compare this figure with Figure 2.22 (the mixture of the same distributions).

An attractive property of the competing risks model is that it can have the U-shaped failure rate, which is based on its additive failure rate (2.112). In the considered case of two competing failure rates, in order to have the U-shaped failure rate for the competing risks model, the FM (risk) with a shorter time to failure should have a decreasing failure rate (DFR), and the second FM should have an IFR. This is illustrated by the following model.

We are considering the competing risks model in which the first distribution is the Weibull one with the scale parameter α equal to 1 and the shape parameter β equal to 0.5 (i.e., DFR distribution). The second distribution is also a Weibull one with the scale parameter α equal to 5 and the shape

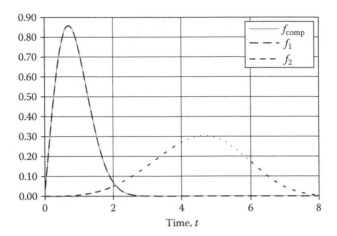

FIGURE 2.25
PDFs of two competing FM. The first FM TTF distribution is a Weibull distribution with scale parameter α equal to 1 and shape parameter of 2. The second distribution is also a Weibull one with scale parameter α equal to 5 and shape parameter of 4. On the graph, the competing risks PDF $f_{comp}(t)$ coincides with PDF of first FM $F_1(t)$.

parameter β equal to 4 (i.e., IFR distribution). The failure rate of this competing failure mode (risks) model is illustrated by Figure 2.26.

Figure 2.27 shows the competing failure λ_{comp} from Figure 2.26 in the logarithmic base 10 scale. The reader might find it interesting to compare the shape of this curve with the shape of the bathtub curve for human mortality (Gavrilov and Gavrilova, 2004) depicted in Figure 2.28.

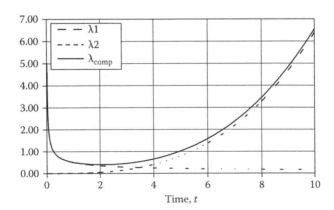

FIGURE 2.26
Failure rate function of two competing FM. The first FM TTF distribution is a Weibull distribution with scale parameter α equal to 1 and shape parameter of 0.5 (i.e., DFR distribution). The second distribution is also a Weibull one with scale parameter α equal to 5 and shape parameter of 4 (i.e., IFR distribution).

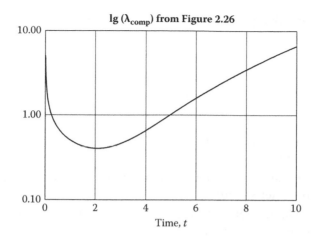

FIGURE 2.27
The competing failure λ_{comp} from Figure 2.26 shown in the logarithmic base 10 scale.

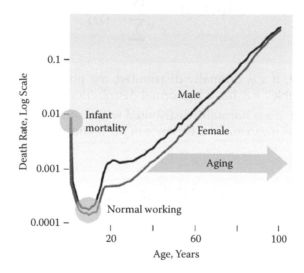

FIGURE 2.28
The bathtub curve for human mortality (failure rate) (Gavrilov and Gavrilova, 2004).

Exercises

1. Show that if the Poisson mean M is an integer, the respective PDFs at M and $M - 1$ are equal, i.e., $Pr(M; M) = Pr(M - 1; M)$ for any integer $M \geq 1$.

2. Random shocks arrive according to the HPP. Assume that only the kth random shock is lethal for the object (component) of interest. Show that the component TTF PDF is the gamma PDF given by (2.19).

Solution: Taking the derivative of (2.18) with respect to time t, the gamma PDF can be obtained as follows:

$$f(t; k, \lambda) = \lambda e^{-\lambda t} \left(\sum_{i=0}^{k-1} \frac{(\lambda t)^i}{i!} - \frac{1}{\lambda} \sum_{i=0}^{k-1} \frac{\lambda (\lambda t)^{i-1}}{(i-1)!} \right) = \lambda e^{-\lambda t} \frac{(\lambda t)^{k-1}}{(k-1)!} \qquad (2.19)$$

3. Show that the mean of the exponential distribution is equal to its scale parameter.

4. Show that the variance of the exponential distribution is equal to its scale parameter squared.

5. Using the failure rate definition (1.4), show that the gamma distribution failure rate is given by

$$h(t; k, \lambda) = \frac{\dfrac{\lambda e^{-\lambda t} (\lambda t)^{k-1}}{(k-1)!}}{e^{-\lambda t} \displaystyle\sum_{i=0}^{k-1} \frac{(\lambda t)^i}{i!}} \qquad (2.20)$$

6. Show that, if x is normally distributed, the positively defined random variable $t = e^x$ has a lognormal distribution.

7. Show that, if x is normally distributed with a mean μ and a variance σ^2, then the mean and the variance of the lognormal distribution are given by

$$\text{MTTF} = e^{\mu + \frac{\sigma^2}{2}} \qquad (2.35)$$

$$\text{Var} = e^{2\mu + \sigma^2} \left(e^{\sigma^2} - 1 \right) = \left[\text{MTTF}(t) \right]^2 \left(e^{\sigma^2} - 1 \right) \qquad (2.36)$$

8. Express the CDF of the lognormal distribution (2.37) in terms of the CDF of the standard normal distribution (2.29).

   Solution: Substituting $s = \dfrac{\ln(\tau) - \mu}{\sigma}$ and $ds = \dfrac{d\tau}{\sigma \tau}$, the lognormal CDF can be written as

$$F(t) = \int_0^{\frac{\ln(t) - \mu}{\sigma}} \frac{1}{\sqrt{2\pi}} \exp\left(-\frac{s^2}{2} \right) ds = \Phi\left(\frac{\ln(t) - \mu}{\sigma} \right)$$

9. Show that the Weibull distribution PDF is given by

$$f(t; \alpha, \beta) = \frac{\beta}{\alpha}\left(\frac{t}{\alpha}\right)^{\beta-1} \exp\left(-\left(\frac{t}{\alpha}\right)^{\beta}\right)$$

10. Show that the Weibull distribution failure rate is given by (2.46)

$$h(t; \alpha, \beta) = \frac{\beta}{\alpha}\left(\frac{t}{\alpha}\right)^{\beta-1} \tag{2.46}$$

11. The so-called *parallel system* of two components is considered. By definition, a parallel system of n components functions if at least one of its components functions. Each of two components of our two-component system has exponentially distributed time to failure. The reliability function of the system $R(t)$ is

$$R(t) = 1 - (1 - \exp(-\lambda_1 t))(1 - \exp(-\lambda_2 t))$$

Derive the respective expressions for the system failure rate and average failure rate.

12. Prove that the variance of the logarithm of TTF stays constant at any stress condition in the domain where a given AL model holds.

13. Show that the coefficient of variation (the ratio of the standard deviation to the mean) stays constant for any stress level, at which a given accelerated life model (2.67) is adequate:

$$CoV = \frac{\sigma(\tau)}{E(\tau)} = const \tag{2.70}$$

14. The PDF at the normal stress condition $f_0(t)$ is the Weibull PDF with the scale parameter equal to 1 and the shape parameter equal to 2. The PDF at the accelerated stress condition $f_1(t)$ was evaluated using relationship (2.72), i.e., $f(t; z) = \psi(z)f_0[t\psi(z, A)]$, with $\psi = 2$, i.e., $f_1(t) = 2 f_0(2t)$. Show that $f_1(t)$ is the Weibull PDF with the scale parameter equal to 0.5 and the shape parameter equal to 2.

15. Show that the PH model relationship for the TTF probability density functions (2.80) can be obtained from (2.77) as

$$f(t; z) = \psi(z)f_0(t)R_0(t)^{\psi(z,A)-1}$$

3

Probabilistic Models for Repairable Objects

In Chapter 2, we dealt mostly with the time to failure (TTF) of nonrepairable objects. Considering different damage models resulting in special classes of TTF distributions (increasing failure rate (IFR), decreasing failure rate (DFR), etc.) and particular TTF distributions (such as the exponential distribution and Weibull distribution), we assumed that the damage was done by some random damaging *events*. The occurrence of these events is modeled by some random process (mostly by the homogeneous Poisson process (HPP) and the nonhomogeneous Poisson process (NHPP)). In this chapter, we discuss these processes in detail, concentrating on their properties and considering the damage made by each event. The main question we are going to address is: If the TTF distribution of an object (system) before the damaging event is $F_0(t)$, what is the TTF distribution just after the event? Is it the same distribution $F_0(t)$, or is it another distribution, say, $F_1(t)$? If it is another distribution, how is it related to $F_0(t)$?

In this chapter, we consider different point processes applicable to different repairable/renewable objects. From an engineering standpoint, these random processes are the models related to main repairable systems of reliability notions, such as repairs, spare stocks, maintenance, preventive maintenance, optimal preventive maintenance, and availability.

In the end, we introduce the so-called Gini-type index for aging/ rejuvenating processes, similar to the one that was introduced in Chapter 2 for nonrepairable objects. In the context of the current chapter, this index provides a universal measure showing how fast a given repairable object ages or rejuvenates. Please note that sometimes we use the words *system* and *object* interchangeably.

3.1 Point Processes as Model for Repairable Systems Failure Processes

In the following discussion, the term *repairable object* is used as a synonym for any repairable or renewable item, e.g., a system, subsystem, or component (part). The term *repair (restoration)* can be applied to the ideal of replacing an old object with a new one, equivalent to a perfect repair (*as-good-as-new* restoration type), and it also can be applied to other types of restoration or fixes as well as to design/manufacturing changes (as in reliability growth modeling (Crow, 1974, 1982)).

In analyzing repairable systems, it is often assumed that the time to repair (or maintenance time) is negligible compared with the mean time between (successive) failures (MTBF). This assumption makes it possible to apply different *point processes* as appropriate models for real-life failure processes. For example, a minor automobile repair might take only a few hours to perform, which is practically negligible compared to the respective MTBF. In this case, the assumption is rather realistic. On the other hand, an opposite example could be the time needed to repair an aviation system, which might be comparable with its predicted MTBF for the so-called soft (noncritical) failures. If a monarchy is considered as a repairable (or *renewable*) system, the medieval public statement "The King is dead, long live the King!" (*Le Roi est mort, vive le Roi!*) shows that the monarchy power transition occurred instantaneously upon the moment of death of the previous monarch, such that the respective event sequence can be modeled by a point process.

It is worth mentioning that the applications of the point processes are not limited to repairable engineering systems. Typical applications of these models can be divided into two groups. The first group includes cases in which undesirable events (incidents) are associated with an internal failure process, for example, a failure process of different engineering systems (ship or nuclear power plant pump failures, or automobile or electrical appliance failures) due to inherited defects such as microcracks in solids or inherited diseases (in humans) or to aging processes. The second group includes cases in which undesirable events are associated with external phenomena, such as strong winds, floods, or earthquakes.

Note that in contrast to repairable objects (systems), any typical reliability model of *nonrepairable* identical units is based on a positively defined *random variable*, such as the time to the first (and the only) failure or the number of cycles to the first failure. In this case, the lifetime of the unit is treated as the realization of an identical and independently distributed (IID) random variable. In a sense, the time to failure as a random variable with a given distribution function can be considered as a special case of the failure process, in which the process is terminated just after the first failure.

Observations of repairable systems are typically the times between successive failures: the time to the first failure, the time between the first and second failures, the time between the second and third failures, and so on. Observations of a point process as a series of successive events that occurred during a time interval are referred to as *realizations* of a given point process. Often, the assumption that these successive failure times are independent and identically distributed (IID) random variables turns out not to be applicable, and different point process models are applied as an appropriate alternative approach to modeling the repairable system failure processes.

It should be noted that the assumption about negligible time to repair might not be critical for some problems related to repairable systems. Let's consider a series of failure-repair events in which each failure is *instantaneously* followed by a repair action. The respective data are the times between failures and repair action durations, which can be represented by the follow-ing series:

$$t_{01}, \tau_1, t_{12}, \tau_2, \ldots, t_{k-1\, k}, \tau_k, \ldots, t_{n-1\, n}, \tau_n \qquad (3.1)$$

where t_{01} is time to the first failure, $t_{k-1\, k}$ is the time between $(k-1)$th and kth failures, and τ_k is the repair time just after the kth failure ($k = 1, 2, \ldots, n$).

In the framework of the repairable system analysis, we can analyze the whole data set (3.1) using the so-called *alternating renewal process*, which is closely related to the notion of system availability. On the other hand, delet-ing the repair times from the data set (3.1), we obtain the following series of times between successive failures, including the time to the first failure:

$$t_{10}, t_{12}, \ldots, t_{k-1\, k}, \ldots, t_{n-1\, n} \qquad (3.2)$$

This data set is typically needed for solving different reliability and risk analysis problems related to repairable systems, such as estimating whether the system is improving, deteriorating, or revealing a constant failure rate, predicting the number of failures observed in a given time interval, develop-ing the respective logistics, or analyzing reliability growth.

Similarly, picking the repair times out of the data set (3.1), we obtain the following series of successive repair times:

$$\tau_1, \tau_2, \ldots, \tau_k, \ldots, \tau_n \qquad (3.3)$$

Based on this type of data, it is possible to determine whether the dura-tion of repair actions of interest are decreasing or not. In other words, we can perform a trend analysis of the repair time similar to the reliability trend analysis.

The notion of the point process introduced below plays the same role in the reliability of repairable objects (systems) as the notion of time-to-failure distribution plays in the reliability of nonrepairable objects (components).

A *point process* can be informally defined as a mathematical model for highly localized events distributed randomly in time. The major random variable of interest related to such processes is the number of failures (or, generally speaking, *events*) $N(t)$ observed in the time interval $(0, t]$, which is why such processes are also referred to as *counting processes*. Using the nondecreasing integer-valued function $N(t)$, the point process $\{N(t), t \geq 0\}$ is defined as the one satisfying the following conditions:

1. $N(t) \geq 0$.
2. $N(0) = 0$.
3. If $t_2 > t_1$, then $N(t_2) \geq N(t_1)$.
4. If $t_2 > t_1$, then $[N(t_2) - N(t_1)]$ is the number of events (e.g., failures) that occurred in the interval $(t_1, t_2]$.

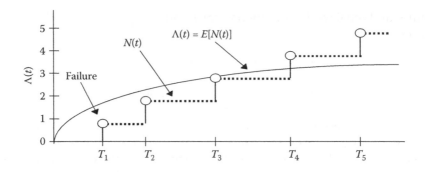

FIGURE 3.1
Geometric interpretation of $N(t)$ and $\Lambda(t)$ for a repairable object.

It is important to note that for most of the point processes considered below, the origin of time, counting $t = 0$, is the moment when the system starts functioning and its age is equal to zero. Note that if the point process is terminated at the time of the first failure, the process becomes the model of the time to failure of a nonrepairable object.

A *trajectory* (*sample path*) or *realization* of a point process is the set of successive failure times of an item: $T_1, T_2, \ldots, T_k \ldots$. It is expressed in terms of the integer-valued function $N(t)$, that is, the number of events observed in the time interval $(0, t]$, as illustrated by Figure 3.1.

$$N(t) = \max(k \,|\, T_k \leq t) \tag{3.4}$$

Simply speaking, (3.4) is the number of events in the interval $(0, t]$.

It is clear that $N(t)$ is a random function. The mean value $E(N(t))$ of the number of failures $N(t)$ observed in the time interval $(0, t]$ is called the *cumulative intensity function* (CIF), the *mean cumulative function* (MCF), or the *renewal function*. In the following, the term *CIF* is used. The CIF is usually denoted by W, that is,

$$W(t) = E(N(t)) \tag{3.5}$$

It should be kept in mind that $N(t)$ given by (3.4) is the *natural estimate* of the CIF $W(t)$. This is the estimate used at the end of this section.

Similar to the cumulative distribution function (CDF) of a random variable, from now on we will assume that CIF $W(t)$ is related to one object (system). The object is supposed to be a member of a population (finite or infinite) consisting of identical (from a reliability standpoint) objects (systems).

Let's recall our statement that the time to failure as a random variable with a given distribution function can be considered as a special case of the point failure process, in which the process is terminated just after the first failure.

Considering such a process, we can see that the CIF of this process is equal to the time-to-failure cumulative distribution function $F(t)$, i.e.,

$$E(N(t)) = F(t)$$

Another important characteristic of point processes is the *rate of occurrence of failures* (ROCOF), which is defined as the derivative of the CIF with respect to time. That is,

$$w(t) = \frac{dW(t)}{dt} \tag{3.6}$$

Based on the above definition of ROCOF, the CIF is sometimes called the *cumulative* ROCOF.

Most of the processes discussed below have monotone ROCOF. The object modeled by a point process with an increasing ROCOF is called an *aging* (or *degrading*) object. Analogously, the object modeled by a point process (PP) with a decreasing ROCOF is called a *rejuvenating* (or *improving*) object. A PP with an increasing (decreasing) ROCOF is called an aging (rejuvenating) point process. It is clear that the CIF of an aging (rejuvenating) point process is concave upward (downward). Later in this chapter, we will discuss different CIF and ROCOF plots of aging and rejuvenating objects (see, e.g., Figures 3.10 and 3.11).

The distribution of time to the first failure of a point process is called the *underlying distribution*. For some point processes this distribution coincides with the distribution of time between successive failures (which is also called the *interarrival time*); for others it does not. The underlying distribution is included in the definition of any particular point process used as a model for the failure/ repair process of repairable systems. The underlying distribution is, in a sense, a nonrepairable component of the respective point process definition.

At this point, it is rational to introduce the notions of *short-term* and *long-term behavior* for point processes (Kaminskiy, 2004). *Short-term behavior* implies that a process is observed during an interval limited by a time close to the mean (or the median) of the respective underlying distribution, i.e., for the time intervals for which the probability of the second and successive failures is much lower than the probability of the first failure. As opposed to short-term behavior, *long-term behavior* is the behavior of a failure process observed during a time much longer than the mean or median of its underlying distribution. Another way to define long-term behavior is to define the observation time interval long enough to successfully apply asymptotic theorems, which are discussed later.

Wrapping up this section, we would like to mention once more that the repairable object models are not restricted only to engineering problems. For example, Rigdon and Basu (2000) model some earthquake processes by applying the point failure-repair process models.

EXAMPLE 3.1
Roman Empire as Renewable System

In the beginning of this section, we mentioned that in the Middle Ages, people took care of satisfying the point process assumptions during the times of transition in monarchy power ("The King is dead, long live the King!"). Now we are going to introduce some well-known data related to the power transitions in the Roman Empire during the times of its decline.

Each power transition (due to whatever reason) from one emperor to another will be considered as a (soft) failure of the system known as the Roman Empire, which is discussed here in terms of repairable (or *renewable*) systems. In this system, a change of an emperor is treated as an event in a failure-repair point process.

The data source we are going to use is the well-known classic *History of the Decline and Fall of the Roman Empire* by Edward Gibbon. The first six-volume edition of this monograph was published between 1776 and 1788. A latest facsimile of this book was published in the Elibron Classics series in 2005.

There is no consensus among historians about the time frame of the decline of the Roman Empire. We are going to analyze only the time interval beginning with Augustus (31 B.C.) through Aurelian (275 A.D.), which includes the following three dynasties from the Gibbon's table of Roman emperors: the Julio-Claudians, the Flavians and Antonines, and the Severans. The related data from the Gibbon's monograph are given in Table 3.1. Using these data, we are going to analyze the cumulative intensity function (3.5), using its simple estimate (3.4).

The estimate (3.4) is not difficult to calculate for any time t equal to a power transition year. In this case our CIF estimate $N(t)$, i.e., cumulative number of power transitions observed up to time t, is equal to the power transition order number, which is given in the first column of Table 3.1.

The plot of $N(t)$ looks concave upward (Figure 3.2), indicating that the Roman Empire can be considered an aging system during the time interval of interest. This is definitely true if the time interval is reduced to (100, 275). Our plot has an interesting cluster of four points in the year 69, which is the year of power transition from the dynasty of Julio-Claudians to the dynasty of Flavians and Antonines. In this year, there were three successive power transitions from one emperor to another (Galba → Otho → Vitelius → Vespasian), which is why this year is called the *year of the four emperors* (Wellesley, 1976). Similarly, A.D. 193 is the year of power transition from the dynasty of the Flavians and Antonines to the Severance dynasty. Despite the fact that there were only two power transitions in 193 (Pertinax → Didius Julianus → Sepimius Severus), this year is referred to as the *year of the five emperors* because there were five claimants for the title of Roman emperor. These five were Pertinax, Didius Julianus, Pescennius Niger, Clodius Albinus, and Septimius Severus (Herodian, 1961).

TABLE 3.1

Roman Emperors between 31 BC and 275 AD

	Dynasty	Emperor	From	Until
1	Julio-Claudians	Augustus	31 BC	14 AD
2	Julio-Claudians	Tiberius	14	37
3	Julio-Claudians	Gaius (Caligula)	37	41
4	Julio-Claudians	Claudius	41	54
5	Julio-Claudians	Nero	54	68
6	Julio-Claudians	Galba	68	69
7	Julio-Claudians	Otho	69	
8	Julio-Claudians	Vitellius	69	
9	Flavians and Antonines	Vespasian	69	79
10	Flavians and Antonines	Titus	79	81
11	Flavians and Antonines	Domitian	81	96
12	Flavians and Antonines	Nerva	96	98
13	Flavians and Antonines	Trajan	98	117
14	Flavians and Antonines	Hadrian	117	138
15	Flavians and Antonines	Antoninus Pius	138	161
16	Flavians and Antonines	Marcus Aurelius	161	180
17	Flavians and Antonines	Commodus	180	192
18	Flavians and Antonines	Pertinax	193	
19	Flavians and Antonines	Didius Julianus	193	
20	Severans	Sepimius Severus	193	211
21	Severans	Aurelius Antoninus (Caracalla)	211	217
22	Severans	Macrinus	217	218
23	Severans	Elagabalus	218	222
24	Severans	Severus Alexander	222	235
25	Severans	Maximinus	235	238
26	Severans	Gordians I and II	238	
27	Severans	Balbinus and Pupienus	238	
28	Severans	Gordian III	238	244
29	Severans	Philip	244	249
30	Severans	Decius	249	251
31	Severans	Trebonianus Gallus	251	253
32	Severans	Aemilianus	253	
33	Severans	Valerian	253	260
34	Severans	Gallienus (253–260 with Valerian)	253	268
35	Severans	Claudis II (Gothicus)	268	270
36	Severans	Quintillus	270	
37	Severans	Aurelian	270	275

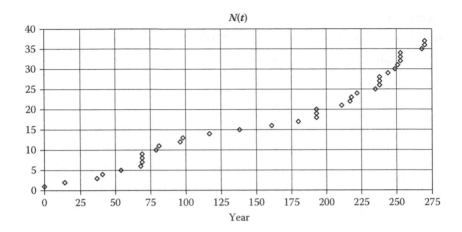

FIGURE 3.2
Number of power transitions in the Roman Empire from Augustus (31 BC) through Aurelian (275 AD).

3.2 Homogeneous Poisson Process as a Simplest Failure-Repair Model

Now we need to introduce some other terms related to the point processes. A point process is said to have *independent increments* if the numbers of events (failures) in mutually exclusive intervals are independent random variables. A point process is called *stationary* if the distribution of the number of events in any time interval depends on the length of the time interval only.

The *homogeneous Poisson process* (HPP) is probably the oldest as well as the simplest *failure-repair* process model. We have already begun discussing this process in Chapter 2 (Section 2.1) as the simplest *shock model*. Now, we are going to talk about the HPP as a failure-repair model in terms of the point processes.

A point process having independent increments is called HPP with parameter $\lambda > 0$ if the number of events (e.g., failures) N in any interval of length $t = (t_2 - t_1)$ has the Poisson distribution with mean $E(N) = \lambda t$. That is,

$$\Pr[n(t_2) - n(t_1) = n] = \frac{[\lambda(t_2 - t_1)]^n}{n!} \exp[-\lambda(t_2 - t_1)] \tag{3.7}$$

where $t_2 > t_1 \geq 0$. Note that the location of the time interval on the time axis does not matter.

It is not difficult to show that for the HPP, the times between successive failures are independent, identically distributed random variables having the exponential distribution with parameter λ, which is the failure rate for this exponential distribution.

In terms of the point processes, the parameter λ is the ROCOF. The mean number of failures $W(t)$ observed in interval $(0, t]$, i.e., the CIF of this point process, is

$$W(t) = \int_0^t \lambda(\tau)d\tau \qquad (3.8)$$

The ROCOF of the HPP is obviously constant, so that

$$W(t) = \lambda t \qquad (3.9)$$

Due to the memoryless property of the exponential distribution (discussed in Section 2.1.2), a repairable system's failure behavior modeled by the exponential distribution cannot be expressed in terms of system age. Simply speaking, the failure behavior does not depend on age. Consequently, any preventive maintenance action does not make sense in the framework of the HPP model.

In many situations it is necessary to model the failure behavior of different items *simultaneously* (e.g., the items are put into service or subject to a test at the same time). Such situations can be modeled by the *superposition* of the respective point processes. The superposition of several point processes is the ordered sequence of all failures that occur in any of the individual point processes. The superposition of several HPP processes with parameters $\lambda_1, \lambda_2, ..., \lambda_k$ is the HPP with $\lambda = \lambda_1 + \lambda_2 + ... + \lambda_k$. A well-known example of the superposition is a failure behavior of a series system (introduced in Chapter 2), composed of elements having exponentially distributed failure times (TTF).

3.3 Renewal Process: As-Good-as-New Repair Model

Another important failure process model for reliability, geophysics, and many other applications is the renewal process (RP). The RP can be considered as a generalization of the HPP (which is a particular case of RP) for the case in which the time between successive failures (TBF) distribution (i.e., the underlying distribution) is an arbitrary continuous distribution of a positively defined random variable. In this case, strictly speaking, the number of failures in any interval of length t no longer follows the Poisson distribution.

The time origin on our time axis is important to the RP and other point processes considered below. It is assumed (unless another assumption is made) that the time origin coincides with the time at which a new (perfectly renewed or repaired) unit is put in a functioning state. In other words, it means that at the time origin, the age of the unit is equal to zero.

For any applicable underlying distribution, the RP models the so-called *as-good-as-new* restorations (an equivalent term is *same as new*) or operates under the so-called *perfect repair assumption*. Note that the term *as good as new* (as well as *perfect*) might be misleading in some cases. For example, for the decreasing failure rate (DFR) distributions, a restoration to the *as-good-as-new* condition cannot be called a perfect repair because the unit having the DFR lifetime distribution has its highest failure rate when it is new (i.e., at zero age). In this case, the repair returning our unit to the state corresponding to its zero age is difficult to call *perfect*. As was previously discussed, the notion of age is not applicable for exponential distribution, so any product having the exponential TBF is always as good as new.

The perfect repair assumption is definitely not appropriate for a multi-component system if only a few of the system components are repaired or replaced upon failure. On the other hand, if a failed system (or a separate component) is replaced by an identical new one, the same-as-new repair assumption is quite appropriate for this system (or component).

In the following, it is also assumed that the cumulative distribution function (CDF) of the underlying distribution, $F(t)$, is continuous together with its probability density function (PDF) $f(t)$. The respective cumulative intensity function (CIF), $W(t)$, for the considered RP can be found as a solution of the so-called *renewal equation*,

$$W(t)= F(t)+ \int_0^t F(t-\tau)dW(\tau) \qquad (3.10)$$

The corresponding ROCOF $w(t)$ can be found as a solution of the equation obtained by taking the derivatives of both sides of (3.10) with respect to t. So we find that

$$w(t)= f(t)+ \int_0^t f(t-\tau)w(\tau)d\tau \qquad (3.11)$$

Closed-form solutions of the above integral equations are available only for the exponential underlying distribution (the case of the HPP) and the gamma underlying distribution.

A number of numeric solutions of the renewal equation have been developed. Smith and Leadbetter (1963) found an iterative solution for the case when the underlying distribution is the Weibull one. Another iterative solution with the Weibull-distributed time between failures was given by White

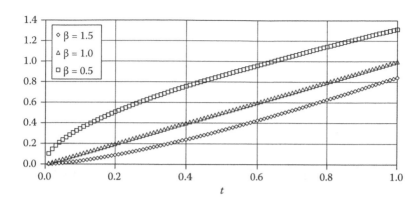

FIGURE 3.3
Cumulative intensity functions of renewal processes with Weibull underlying distributions having scale parameter equal to 1 and different shape parameters β.

(1964). Baxter et al. (1982) offered a numerical integration approach that covers the cases of the following underlying distributions: Weibull, gamma, lognormal, truncated normal, and inverse normal. Garg and Kalagnanam (1998) proposed the Pade approximation approach to solve the renewal equation for the case of the inverse normal underlying distribution. Blischke and Murthy (1994) used a Monte Carlo simulation, which is a universal numerical approach to the problem.

The cumulative intensity functions for the RP with the Weibull underlying distributions having the scale parameter equal to 1 and the different shape parameters are depicted in Figure 3.3.[*] Note that the case β = 1 is the HPP (the exponential underlying distribution). For the case β = 1.5 (IFR underlying distribution), the cumulative intensity function is concave upward, whereas for the case β = 0.5 (the DFR underlying distribution), the cumulative intensity function is concave downward, and for the case β = 1, it is a straight line (HPP). This property corresponds to the increasing (for the case β = 1.5) and decreasing (for the case β = 0.5) ROCOFs of the corresponding RPs. The respective ROCOF plots are given in Figure 3.4. Please note that the upper limit on the time axis in Figures 3.3 and 3.4 is equal to the scale parameter of the underlying distributions (or its 63rd percentile).

Table 3.2 contains the medians and means of the underlying distributions for the RPs illustrated by Figures 3.3 and 3.4. Figures 3.3 and 3.4 illustrate the short-term behavior of the respective RPs. In contrast to these figures, Figure 3.5 illustrates the long-term behavior of the CIFs of the same RPs. The CIFs in Figure 3.5 look like straight lines, implying that the corresponding

---

[*] The data displayed in this figure, as well as in many of the following figures, were obtained using the Monte Carlo simulation.

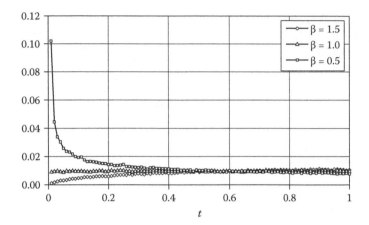

FIGURE 3.4

ROCOF functions of renewal processes with Weibull underlying distributions having scale parameter equal to 1 and different shape parameters β.

TABLE 3.2

Medians and Means of Underlying Distributions of RPs Illustrated by Figures 3.3 and 3.4

Distribution	Parameters	Median	Mean
Weibull	α = 1.0, β = 1.5	0.78	0.90
Weibull	α = 1.0, β = 0.5	0.48	2.00
Weibull (exponential)	α = 1.0, β = 1.0	0.69	1.00

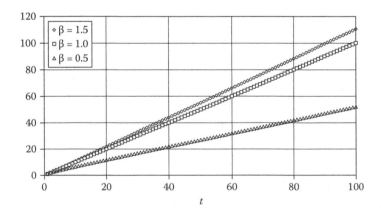

FIGURE 3.5

Long-term behavior of RP. Cumulative intensity functions of renewal processes having the Weibull underlying distributions with scale parameter α = 1 and different shape parameters β.

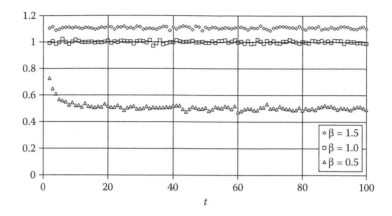

FIGURE 3.6
Long-term behavior of RP. ROCOF functions of renewal processes having the Weibull underlying distributions with scale parameter $\alpha = 1$.

ROCOFs might be constant. The respective ROCOFs are shown in Figure 3.6. The figure reveals that each ROCOF gets closer to a constant limit as time increases. This limit is given by the so-called *Blackwell's theorem*. The theorem states that for an RP with an underlying continuous distribution with mean $E(t)$, the following *limiting* relationship holds:

$$\lim_{t \to \infty}\left(W(t+a) - W(t)\right) = \frac{a}{F(t)} \tag{3.12}$$

where $a > 0$.

The approximate values of the ROCOF functions used in Figure 3.6 may be estimated as

$$w(t) \approx \frac{W(t+a) - W(t)}{a} \tag{3.13}$$

with $a = 1.0$.

The average value of the ROCOF for the RP illustrated by Figure 3.6 and the values of the ROCOF estimated using (3.13) turn out to be close enough.

EXAMPLE 3.2
For the RP with the Weibull underlying distribution having the shape parameter $\beta = 1.5$ and the scale parameter $\alpha = 1$, the average value of the ROCOF is estimated as 1.11 (see Figure 3.6), which practically coincides with the value $1/E(t) = 1/0.90 = 1.11$, based on Equation (3.12) with $a = 1$. In other words, this example shows that the long-term behavior of the RP can be rather accurately described by Blackwell's theorem.

3.4 Nonhomogeneous Poisson Process:
As-Good-as-Old Repair Model

A point process having independent increments is called the nonhomogeneous Poisson process (NHPP) with time-dependent integrable ROCOF $\lambda(t) > 0$, if the probability that exactly n events (e.g., failures) occur in any interval (a, b) has the Poisson distribution with the mean equal to

$$\int_a^b \lambda(t)dt$$

That is,

$$P[N(b) - N(a) = n] = \frac{\left[\int_a^b \lambda(t)dt\right]^n e^{-\int_a^b \lambda(t)dt}}{n!} \tag{3.14}$$

for $n = 0, 1, 2, \ldots, \infty$, and where $N(c)$ is the number of events that occurred in the interval $(0, c)$ and $N(0) = 0$. Opposite to the RP, the times between successive events (e.g., failures) in the framework of the NHPP model are neither independent nor identically distributed.

Similar to the RP discussed in the previous section, it is further assumed that for the NHPP applications, the time origin (or zero age) is the time when a new (renewed or repaired) unit has been put into operation.

Based on the above definition, the CIF and ROCOF of NHPP can be written, respectively, as

$$W(t) = \int_a^b \lambda(\tau)d\tau \tag{3.15}$$

and

$$w(t) = \lambda(t) \tag{3.16}$$

The cumulative distribution function (CDF) of time to the *first* failure (i.e., the CDF of the underlying distribution) for the NHPP can be found as

$$F(t) = 1 - \Pr[N(t) - N(0) = 0] = 1 - \exp(-W(t)) \tag{3.17}$$

where $W(t)$ is the CIF given by (3.15).

Let's consider a series of failures that occur according to the NHPP with ROCOF $\lambda(t)$. Let t_k be the time to the kth failure, so at this moment, the

ROCOF is equal to $\lambda(t_k)$. Using (3.17), the probability that no failure occurs in interval (t_k, t), where $t > t_k$, can be written as

$$R(t_k, t) = e^{-\int_{t_k}^{t} \lambda(\tau)d\tau} = \frac{e^{-\int_{0}^{t} \lambda(\tau)d\tau}}{e^{-\int_{0}^{t_k} \lambda(\tau)d\tau}} = \frac{R(t)}{R(t_k)} \tag{3.18}$$

The above expression is the conditional reliability function of an object (repairable system) having age t_k. In other words, we can consider the NHPP as a process in which each failed object is instantaneously replaced by an identical, functioning one having the same age as the failed one at the failure time. This type of restoration model is called the *same-as-old* (or *minimal repair*) condition.

Another very important property of the NHPP, under the given assumptions, follows from Equation (3.18). If t_k is equal to zero, the probability (3.18) takes the following form:

$$R(t) = \exp\left(-\int_{0}^{t} \lambda(\tau)d\tau\right) \tag{3.19}$$

Comparing (3.19) with the expression for the reliability function (1.5) from Chapter 1, we see that the *ROCOF of the NHPP coincides with the failure (hazard) rate function of the underlying (time to the first failure) distribution.* Respectively, the NHPP cumulative intensity function coincides with the cumulative failure rate of the underlying distribution. In other words, all future failure-repair behavior of a repairable object is completely defined by this distribution (like a gene responsible for the object longevity). It also means that just after any repair/maintenance action carried out at time t, the ROCOF is equal to the current value of the failure rate of the underlying distribution. So, we can also consider the NHPP as a process in which each failed object/system is instantaneously replaced by an identical one having exactly the same failure rate as the failed one.

Is the same-as-old restoration a realistic assumption? The answer depends on the application. Applied to a single-component system, it is definitely not a realistic assumption. For a complex system, composed of many components having close reliability functions, this assumption is much more realistic, because only a small fraction of the system's components is repaired, with a small resultant change of the system failure rate (Hoyland and Rausand, 1994).

There are two possible ways to define a particular NHPP. The first one is to suggest an ROCOF, and then, based on the ROCOF function, to find the respective underlying distribution (the distribution of time to the first failure). It can be based on a kind of preliminary analysis of reliability tests or early field data collected. The second way is just an opposite of the first one: we begin with choosing the underlying distribution, and based on the chosen distribution, we use its failure rate as the corresponding ROCOF. The second approach (Krivtsov, 2007) might result in the ROCOF, which does not have a simple closed form as in the case of the lognormal distribution or a mixture of two distributions.

We begin with two particular cases of the NHPP defined by specific ROCOF functions.

3.4.1 Power Law ROCOF and Weibull Distribution

An important particular case of the NHPP is the case when the ROCOF is a power function of time. Let's write the power function in the following form:

$$w(t) = \frac{\beta}{\alpha}\left(\frac{t}{\alpha}\right)^{\beta-1} \quad t \geq 0, \ \alpha, \beta > 0 \tag{3.20}$$

The above expression is easily recognizable as the failure rate of the Weibull distribution (2.46). Recalling (3.16), we see that the NHPP with ROCOF (3.20) has the Weibull underlying distribution with the scale parameter α and the shape parameter β. We can also consider a shock model, in the framework of which the fatal shocks occur according to the NHPP process with the rate of occurrence of shocks given by (3.20). The process stops just after the first fatal shock. This process, known as the Weibull process, also results in the Weibull time-to-failure distribution.

The NHPP process with the power function ROCOF (3.20) is sometimes referred to as the Weibull NHPP process, the power law NHPP process, or the Crow-AMSAA model. Statistical procedures for this model were developed by Crow (1974, 1982, 1990), based on suggestions of Duane (1964). These procedures can also be found in MIL-HDBK-781 (1996) and IEC International Standard 1164 (1995). The main reliability applications of the power law model are associated with reliability monitoring and reliability managerial problems, which are optimistically called reliability growth.

3.4.2 Log-Linear ROCOF and Truncated Gumbel Distribution

Another important particular case of NHPP is the case when the ROCOF is a log-linear function. That is,

$$w(t) = \exp(\beta_0 + \beta_1 t) \quad t \geq 0 \tag{3.21}$$

This model was proposed and statistically developed by Cox and Lewis (1966), which is why the model is often referred to as the *Cox-Lewis* model.

Similar to the power law model, the log-linear model has a monotonic ROCOF, which can be increasing (if $\beta_1 > 0$), decreasing (if $\beta_1 < 0$), or constant (if $\beta_1 = 0$). The respective underlying (time to the first failure) distribution is the so-called *truncated Gumbel* (smallest extreme value) distribution. This distribution has the following cumulative distribution function:

$$F(t) = 1 - \exp(-\alpha(e^{\beta_1 t} - 1)) \tag{3.22}$$

where

$$\alpha = \frac{\exp(\beta_0)}{\beta_1}$$

and $t \geq 0$.

The physical model behind this distribution can be found in many reliability books, e.g., Lloyd and Lipow (1962), Mann et al. (1974), Kapur and Lamberson (1977), and Hoyland and Rausand (1994, 2004). The failure mechanism is corrosion. Let us assume that we deal with an object, say a pipe, with a thickness D that is subjected to corrosion. A new pipe is supposed to have a certain number n of microscopic pits, and each pit has a random depth d_i, $i = 1, 2, \ldots, n$. The corrosion increases the depth of these pits so that the pipe fails when one of the pits' depth reaches the value equal to D, i.e., when the respective pit becomes a hole. It is assumed that the initial depths d_i ($i = 1, 2, \ldots, n$) are the independent random variables identically distributed according to the truncated exponential distribution at depth D with parameter λ. The respective probability density function of the initial depths is

$$f_d(d) = \frac{\lambda e^{-\lambda d}}{1 - e^{-\lambda D}}, \qquad 0 \leq d \leq D \tag{3.23}$$

and the corresponding cumulative distribution function is given by

$$F_d(d) = \Pr(d_i < d) = \frac{1 - e^{-\lambda d}}{1 - e^{-\lambda D}} \tag{3.24}$$

Let t_i denote the time to failure related to the growth of the ith pit. Further, it is assumed that the time to failure t_i is proportional to the residual thickness, i.e., $t_i = k (D - d_i)$, where k is a material corrosion rate. In this case, the cumulative distribution function of time to failure is

$$F(t) = P_r(t_i \leq t) = P_r\left(d_i \geq D - \frac{t}{k}\right) \tag{3.25}$$

Using (3.24), one can write this function as

$$F(t) = P_r\left(d_i \ge D - \frac{t}{k}\right) = \frac{e^{\frac{\lambda t}{k}} - 1}{e^{\lambda D} - 1}, \qquad 0 \le t \le kD \tag{3.26}$$

The suggested reliability model of the pipe is the series system composed from n components so that the pipe reliability function is $R(t) = (1 - F(t))^n$. Further, it is assumed that the number of pits is very large, i.e., $n \to \infty$, and $F(t)$ is small enough, so that one gets the following approximation:

$$R(t) = \left(1 - F(t)\right)^n \approx e^{-nF(t)} \tag{3.27}$$

Applying expression (3.26), it can be written as

$$R(t) \approx e^{-n\frac{e^{\frac{\lambda t}{k}} - 1}{e^{\lambda D} - 1}} \tag{3.28}$$

Introducing a new parameter $\alpha = n/(e^{\lambda D} - 1)$ and $\beta_1 = -\lambda/k$, the above reliability function is rewritten as

$$R(t) \approx \exp\left(-\alpha\left(e^{\beta_1 t} - 1\right)\right) \tag{3.29}$$

This is the reliability function of the truncated Gumbel distribution, the CDF of which is given by (3.22). Thus, if one chooses to apply the NHPP model with the log-linear ROCOF, it means that the underlying distribution of this NHPP will be the truncated Gumbel distribution.

3.4.3 Linear ROCOF and Competing Risks

The NHPP with the simple linear ROCOF, i.e.,

$$w(t) = \beta_0 + \beta_1 t \qquad t \ge 0 \tag{3.30}$$

is not as popular as the models discussed in the previous section. This might be explained by the fact that for any time interval of interest $[0, t]$, the parameters of ROCOF (3.30) must obviously satisfy the following inequality:

$$\beta_0 + \beta_1 t \ge 0 \tag{3.31}$$

If both parameters are positive, the respective underlying distribution is the TTF distribution of the series system (competing risks model), as it is composed of a component with the exponential TTF distribution and a component with the Rayleigh distribution, which is the Weibull distribution with a shape parameter equal to 2. This underlying distribution might look too artificial. However, we can apply this model, considering it as the two-term Maclaurin series approximation of an unknown failure rate function defining our underlying distribution.

The risk analysis applications of the linear model (3.31) are discussed by Vesely (1991) and Atwood (1992).

Statistical data analyses for the above simple NHPP models are well developed and can be found in many books on reliability data analysis, e.g., Cox and Lewis (1966), Crowder et al. (1991), Rigdon and Basu (2000), and Nelson (2003).

3.4.4 NHPP with Nonmonotonic ROCOF

As was mentioned earlier, some ROCOF functions might result in underlying distributions that cannot be expressed in simple closed forms. The ROCOF functions considered in Sections 3.4.1 to 3.4.3 are simple monotonic functions. In this section, we consider an NHPP model having a *nonmonotonic* rate of occurrence.[*] The model was developed by Lee, Wilson, and Crawford (1991). It can be applied to the situations in which the rate of occurrence exhibits locally cyclic behavior as well as a long-term evolutionary trend. It was applied to the storm arrival process observed at an offshore drilling site in the Arctic Sea.

The events (simply called *storms*) considered in Lee et al. (1991) are arrivals of sea waves with heights greater than 20 ft. The observed waves exhibit seasonal variations corresponding to the meteorological conditions at a site. Because the rate of occurrence of storms varies between seasons and from year to year (see Figure 3.7), successive storms are assumed to be statistically interdependent, so the NHPP model might be appropriate.

The suggested rate of occurrence of storms is given by the following exponential rate function:

$$w(t) = \exp\left(\sum_{i=0}^{m} a_i t^i + \gamma \sin(\omega t + \phi) \right) \tag{3.32}$$

where

$$\sum_{i=0}^{m} a_i t^i$$

is a polynomial function representing a general trend over time, and $\gamma \sin(\omega t + \phi)$ is a periodic function representing a cyclic effect exhibited by the process. The function (3.32) is called the exponential-polynomial-trigonometric function (EPTF).

---

[*] The events considered here are storms, not failures, and so we do not use the term *ROCOF* in this section.

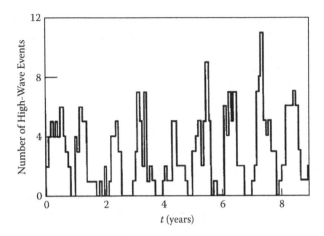

FIGURE 3.7
Number of high-wave events (storms) observed each month over a 9-year observation interval at a site in the Arctic Sea. (From Lee, S., Wilson, J. R., and Crawford, M. M., *Commun. Stat. Simul. Comput.*, 20(2–3), 777–809, 1991.)

The respective underlying distribution for such a complicated model is difficult to obtain in a closed form. Using (3.17) we can, at least, write the CDF of the underlying distribution as

$$F(t) = 1 - \exp\left(-W(t)\right)$$

$$= 1 - \exp\left(-\int_0^t \exp\left(\sum_{i=0}^m a_i \tau^i + \gamma \sin(\omega\tau + \phi)\right) d\tau\right) \tag{3.33}$$

which can be evaluated numerically.

Lee, Wilson, and Crawford (1991) developed a numerical approach to the estimation of parameters of (3.33), as well as an approach to the estimation of the degree of the polynomial component. For their case study, model (3.32) was applied with the degree $m = 3$. The respective fitted and estimated (empirical) cumulative intensity functions are shown in Figure 3.8. It reveals a nonmonotonic rate of occurrence of the storms and a good model fit.

3.5 Generalized Renewal Process

In the previous section, we mentioned some weaknesses related to the NHPP as the minimal repair model. The validity of the assumed minimal repair condition (NHPP) and the perfect repair condition (as the one used

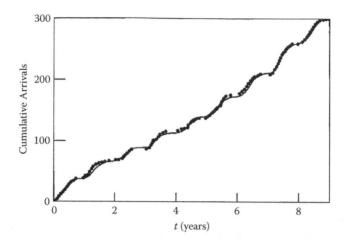

FIGURE 3.8

Empirical cumulative intensity function, i.e., cumulative number of storms (dotted curve), and fitted cumulative intensity function (solid curve). (From Lee, S., Wilson, J. R., and Crawford, M. M., *Commun. Stat. Simul. Comput.*, 20(2–3), 777–809, 1991.)

in the definition of the renewal process (RP)) as realistic assumptions have been criticized by many authors, for example, Ascher and Feingold (1984), Lindqvist (1999), and Thompson (1981). Thompson notes that the NHPP "is a non-intuitive fact that is casting doubt on the NHPP as a realistic model for repairable systems. Use of an NHPP model implies that if we are able to estimate the failure rate of the time to the first failure, such as for a specific type of automobiles, we at the same time have an estimate of the ROCOF of the entire life of the automobile." On the other hand, Ascher and Feingold, discussing the RP, point out, "If an automobile were modeled by a renewal process, its age at any instant in time would be the backward recurrence time (mileage) to the most recent failure/repair!" And finally, Lindqvist notes: "For many applications it is more reasonable to model the repair action by something in between the two given extremes (RP & NHPP)."

Nevertheless, for many years, these processes were the most commonly used models for the failure process. As mentioned above, the RP can be used to model the situations with restoration to an as-good-as-new state (perfect repair assumption); meanwhile, the NHPP is applied to the situations with the same-as-old restoration (minimal repair assumption). In a sense, these two assumptions can be considered extreme, from both theoretical and practical standpoints. In order to avoid this extremism, several generalizing models have been introduced in recent years. References include Brown and Proschan (1982), Kijima and Sumita (1986), Zhang and Jardine (1998), and Lindqvist (1999).

Among these models, the generalized renewal process (GRP) introduced by Kijima and Sumita (1986) is the most attractive one, since it covers not only the RP and the NHPP, but the intermediate younger-than-old-but-older-than-new

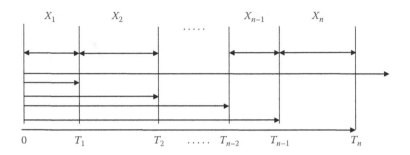

FIGURE 3.9
Times between failures X_i and times to failure T_i; $T_0 \equiv 0$.

repair assumptions.[*] The introduced GRP results in the so-called G-renewal equation, which is a generalization of the ordinary renewal equation (3.10).

It should be noted that as a rather new model, the GRP model has been used in many applications, including the automobile industry (Kaminskiy and Krivtsov, 2000a,b), oil industry (Hurtado, Joglar, and Modarres, 2005), reliability studies of hydroelectric power plants (Kahle, 2005), and electrical power supply systems (Jiménez and Villalón, 2006). Veber, Nagode, and Fajdiga (2008) considered a GRP model with the underlying Weibull mixture distribution.

The GRP model is based on the notion of *virtual age*, which is introduced as follows. Let A_n denote the virtual age of a system immediately after the nth repair. If $A_n = y$, then the system has time to the $(n + 1)$th failure X_{n+1} (see Figure 3.9), which is distributed according to the following cumulative distribution function (CDF):

$$F\left(X|A_n = y\right) = \frac{F(X+y) - F(y)}{1 - F(y)} \tag{3.34}$$

where $F(X)$ is the CDF of the time to the first failure distribution of the system when it was new (the underlying distribution).

The CDF (3.34) is the conditional CDF of the system at age y.

The *real age* of the system obviously is the following sum of the times between successive failures:

$$S_n = \sum_{i=1}^{n} X_i \tag{3.35}$$

with $S_0 = 0$ and X_1 equal to the time to the first failure.

---

[*] Kijima and Sumita called their process G-renewal. Later on, the process was interchangeably referred to as the *general* and *generalized* renewal process by other authors.

In the framework of the considered GRP model, it is assumed that the nth repair can affect the damage incurred only during the time between the $(n-1)$th and nth failures (i.e., in the latest period of system operation), so that the respective virtual age after the nth repair is

$$A_n = A_{n-1} + q\,X_n = q\,S_n, \qquad n = 1, 2, \ldots \tag{3.36}$$

where q is the so-called *parameter of rejuvenation* (or *repair effectiveness parameter*) and the virtual age of a new system $A_0 = 0$, so that according to (3.34), the time to the first failure has the CDF $F(t|0) \equiv F(t)$, which is the underlying distribution. The GRP with restoration condition (3.36) is known as the *Kijima model I*.

It is clear that this restoration condition assumes that the nth repair (restoration) affects only the X_n component of the virtual age. It is also clear that when $q = 0$, this process coincides with the ordinary renewal process (RP), thus modeling the same-as-new repair assumption. When $q = 1$, the system is restored to the same-as-old condition, which is the case of the non-homogeneous Poisson process (NHPP). The case of $0 < q < 1$ falls between the same-as-new (RP) and same-as-old (NHPP) repair assumptions. Finally, when $q > 1$, the virtual age $A_n > S_n$. In this case, the repair further damages the system (if the failure process has an increasing ROCOF) to a higher degree than it was just before the respective failure, or the repair improves the system (in the case of the failure process having a decreasing ROCOF), which corresponds to the older-than-it-was repair assumption (Kaminskiy and Krivtsov, 1998, 2000a,b). It is important to note that the above interpretation is good for the system with monotone (increasing or decreasing) ROCOF.

In the framework of the so-called *Kijima model II*, the restoration condition is

$$A_n = q(A_{n-1} + X_n) = q\,(q^{n-1}\,X_1 + q^{n-2}\,X_2 + \ldots + X_n) \qquad n = 1, 2, \ldots \tag{3.37}$$

The choice between the two Kijima models can be mainly based on the physics of failure considerations. The following discussion is limited to the Kijima model I.

The expected number of failures in an interval $(0, t)$, i.e., the cumulative intensity function (CIF) $W(t)$ introduced in Section 3.1, is given by a solution of the so-called G-renewal equation (Kijima and Sumita, 1986):

$$W(t) = \int_0^t \left(g(\tau|0) + \int_0^\tau h(x)g(\tau - x|x)dx \right) d\tau \tag{3.38}$$

where

$$g(t|x) = \frac{f(t+qx)}{1 - F(qx)}, \qquad t, x \geq 0$$

$F(t)$ and $f(t)$ are, respectively, the CDF and PDF of the time to the first failure (underlying) distribution, and $g(t|x)$ is a conditional probability function such that $g(t|0) = f(t)$. Compared to the ordinary renewal equation (3.10), the G-renewal equation (3.38) has one additional parameter, which is the repair effectiveness parameter q.

By definition, the rate of occurrence of failures (ROCOF) is $w(t) = d(W(t))/dt$, so one gets the following equation for the ROCOF of the GRP:

$$w(t) = g(t|0) + \int_0^t w(x)g(t-x|x)dx \qquad (3.39)$$

Kijima and Sumita (1986) showed that the Volterra integral of (3.39) has a unique solution. It should be noted that the closed-form solutions of Equations (3.38) and (3.39), and even numerical solutions, are difficult to obtain, since each equation contains a recurrent infinite system (Finkelstein, 1997). A Monte Carlo–based solution is, however, possible and is discussed by Kaminskiy and Krivtsov (1998).

Kijima, Morimura, and Suzuki (1988) point out that the numerical solution of the G-renewal equation is very difficult in the case of the Weibull underlying distribution. This statement is not very true in the situations where the Monte Carlo approach is applied, as is illustrated by many examples discussed below. All of them were developed using the Monte Carlo approach.

Similar to Section 3.3 (*renewal process*), below we will discuss the short-term and the long-term behavior of the GRP.

Figure 3.10 illustrates the *short-term behavior* of GRP CIFs having the aging (IFR) Weibull underlying distribution (with scale parameter $\alpha = 1$ and shape parameter $\beta = 1.5$ and different rejuvenation parameters q, including $q = 0$, which corresponds to the RP, and $q = 1$, corresponding to the NHPP. For this case, the NHPP CIF is the upper bound and the RP CIF is the lower bound for any GRP CIF with $0 < q < 1$.

Figure 3.11 shows the short-term behavior of the respective GRP ROCOF. Similar to Figure 3.10, the NHPP ($q = 1$) exhibits the highest ROCOF, and the RP ($q = 0$) the lowest.

Figures 3.12 and 3.13 are similar to Figures 3.10 and 3.11, except for the underlying distributions, which are the DFRs for the processes depicted in Figures 3.12 and 3.13. For the GRPs with DFR underlying distributions (in contrast to the GRP with IFR underlying distributions), the NHPP CIF is the lower bound and the RP CIF is the upper bound for any GRP CIF with rejuvenation parameter $0 < q < 1$.

Figure 3.14 illustrates the *long-term behavior* of GRP CIFs having the aging (IFR) Weibull underlying distribution with scale parameter $\alpha = 1$ and shape parameter $\beta = 1.5$ (IFR distribution) and different rejuvenation parameters q, including $q = 0$, which corresponds to the RP, and $q = 1$, which corresponds to the NHPP.

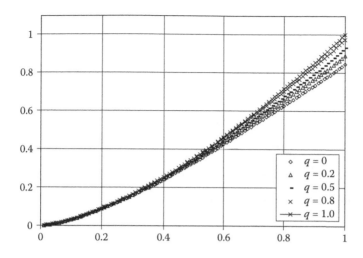

FIGURE 3.10
Short-term behavior of GRP. Cumulative intensity functions of the processes have the aging (IFR) Weibull underlying distribution (with scale parameter $\alpha = 1$ and shape parameter $\beta = 1.5$) and different rejuvenation parameters q.

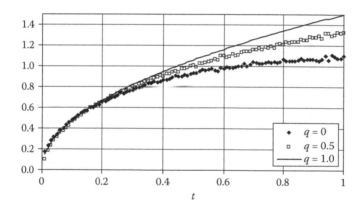

FIGURE 3.11
Short-term behavior of GRP. The ROCOFs of GRPs have the same aging (IFR) Weibull underlying distribution as in Figure 3.10 and different rejuvenation parameters q. The ROCOF of NHPP ($q = 1$) is based on its explicit expression.

As in the case of the short-term behavior, the NHPP CIF is the upper bound and the RP CIF is the lower bound for any GRP CIF with $0 < q < 1$. Figure 3.15 shows the long-term behavior of the respective GRP ROCOFs. Similar to Figure 3.11, the NHPP ($q = 1$) exhibits the highest ROCOF, and the RP ($q = 0$) the lowest one. Note that unlike the RP, the ROCOF of the NHPP and the GRP do not reveal the limiting behavior, similar to the one given by Blackwell's theorem discussed in Section 3.3.

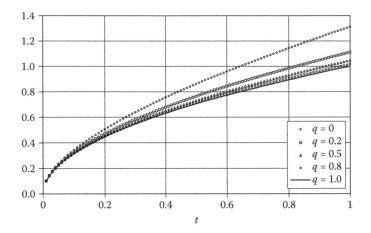

FIGURE 3.12
Short-term behavior of GRP. Cumulative intensity functions of the processes have the same DFR Weibull underlying distribution (with scale parameter $\alpha = 1$ and shape parameter $\beta = 0.5$) and different rejuvenation parameters q.

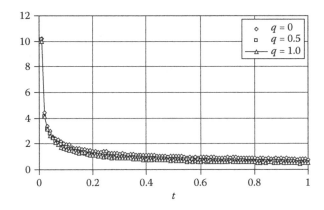

FIGURE 3.13
Short-term behavior of GRP. The ROCOFs of GRPs have the same DFR Weibull underlying distribution as in Figure 3.12 and different rejuvenation parameters q.

Figures 3.16 and 3.17 are similar to Figures 3.14 and 3.15, except for the underlying distributions, which are the DFRs for the processes depicted in Figures 3.16 and 3.17. Note that for the GRPs with DFR underlying distributions (in contrast to the GRP with IFR underlying distributions), the NHPP CIF is the lower bound and the RP CIF is the upper bound for any GRP CIF with rejuvenation parameter $0 < q < 1$.

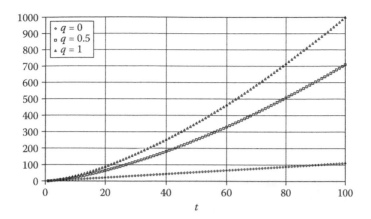

FIGURE 3.14
Long-term behavior of GRP. Cumulative intensity functions of the processes have the aging (IFR) Weibull underlying distribution (with scale parameter $\alpha = 1$ and shape parameter $\beta = 1.5$) and different rejuvenation parameters q.

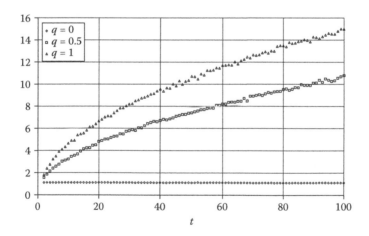

FIGURE 3.15
Long-term behavior of GRP. The ROCOFs of GRPs have the same aging (IFR) Weibull underlying distribution as in Figure 3.14 and different rejuvenation parameters q.

EXAMPLE 3.3
(Kaminskiy and Krivtsov, 2000a,b)

The warranty data collected on a system during the first 18 months (see Table 3.3) were analyzed using the GRP model. The Weibull distribution with the shape parameter β and the scale parameter α was assumed as the underlying distribution. The obtained estimates of GRP parameters are $\beta = 1.8$, $\alpha = 24$, and $q = 0.7$. The solid line in Figure 3.18 represents the fitted GRP cumulative intensity function (CIF).

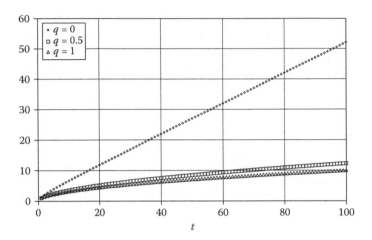

FIGURE 3.16
Long-term behavior of GRP. Cumulative intensity functions of the processes have the DFR Weibull underlying distribution (with scale parameter $\alpha = 1$ and shape parameter $\beta = 0.5$) and different rejuvenation parameters q.

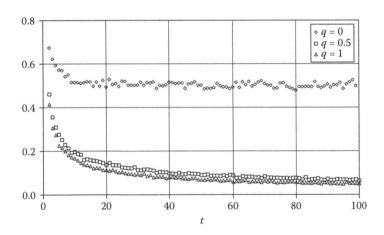

FIGURE 3.17
Long-term behavior of GRP. The ROCOFs of GRP have the same DFR Weibull underlying distribution as in Figure 3.16 and different rejuvenation parameters q.

TABLE 3.3

Warranty Data for a Repairable System (population size, $N = 100,000$)

Month in service, t	3	6	9	12	15	18	21	24	27
Cumulative number of failures per system	0.03	0.09	0.14	0.24	0.38	0.54	0.70	0.90	1.17

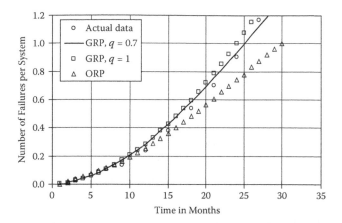

FIGURE 3.18
GRP CIF shows a good fit to Example 3.3 of data in the interval (0, 18] months (used for the GRP CIF estimation), as well as in the interval (18, 30] months (obtained by extrapolation). The figure also shows the extreme repair conditions modeled by the renewal process (GRP with $q = 0$) and the nonhomogeneous Poisson process (GRP with $q = 1$).

The estimated GRP CIF shows a good fit to the data not only in the interval (0, 18] months, which were used for estimation, but also in the remaining interval (18, 30] months (obtained by extrapolation), as shown in Figure 3.18. The figure also shows the extreme repair conditions modeled by the renewal process (i.e., the GRP with $q = 0$) and the nonhomogeneous Poisson process (i.e., the GRP with $q = 1$).

3.6 Inequalities for Reliability Measures and Characteristics for Renewal and Generalized Renewal Processes

From our discussion, it is clear that evaluating the cumulative intensity function (CIF) of the renewal process (RP) and generalized renewal process (GRP) is far from trivial, which is why applying a simple bound on the respective CIFs can be considered a useful shortcut.

Hoyland and Rausand (1994, 2004) give a simple upper bound on the CIF of the RP with the underlying cumulative distribution function $F(t)$ as

$$UB(t) = \frac{F(t)}{1 - F(t)} \tag{3.40}$$

Let's consider a nonhomogeneous Poisson process (NHPP) having the ROCOF $r(t)$, which is, as we know, equal to the failure rate of its underlying distribution, which is denoted by $F(t)$. We assume that $F(t)$ is an increasing failure rate (IFR) distribution. That is,

$$r(t) = \frac{f(t)}{1 - F(t)} \tag{3.41}$$

is an increasing or, strictly speaking, nondecreasing function.

Let $M_{NHPP}(t|F(t))$ be the CIF of the NHPP with ROCOF given by (3.41). That is,

$$M_{NHPP}\big(t|F(t)\big) = \int_0^t r(\tau)d\tau \tag{3.42}$$

$$= -\ln\big(1 - F(t)\big)$$

and let $M_{RP}(t|F(t))$ be the CIF of the RP having the same IFR underlying distribution $F(t)$. In this case, it is clear that the CIF of the NHPP is always greater than the CIF of the RP having the same underlying IFR distribution, $F(t)$. That is,

$$M_{RP}\big(t|F(t)\big) < -\ln\big(1 - F(t)\big) \tag{3.43}$$

$$0 < t < \infty$$

Based on inequality (3.43), one can use the CIF of the introduced NHPP as an upper bound on the CIF of the RP having any underlying IFR distribution $F(t)$. That is,

$$UB(t) = -\ln(1 - F(t)) \tag{3.44}$$

It is easy to show that this upper bound is always more strict than the upper bound (3.40). That is,

$$-\ln\big(1 - F(t)\big) \le \frac{F(t)}{1 - F(t)}, \qquad 0 \le t < \infty \tag{3.45}$$

and therefore should be used instead of (3.40) when the underlying distribution is IFR. The same upper bound (3.45) can be applied to the CIF of the GRP with the IFR underlying distribution and repair effectiveness parameter $q < 1$.

The above-discussed bounds might turn out to be practically valuable only when applied to the short-term observation intervals introduced in Section 3.3. In that section, we also defined the long-term behavior as the behavior of a failure process observed during a time much longer than the mean or median of its underlying distribution. Another possible way to define the long-term behavior is to define the observation time interval long enough to successfully apply the asymptotic theorems.

For a nonformal definition of the long-term observation interval, let us recall the upper bound for RPs introduced by Barlow and Proschan (1996).

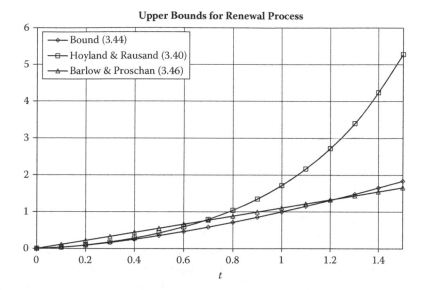

FIGURE 3.19
Upper bounds for the renewal process with the Weibull underlying distribution with scale parameter equal to 1 and shape parameter equal to 1.5.

Their nonparametric (distribution-free) upper bound on the CIF of the RP with an NBUE[*] underlying distribution is given by

$$UB(t) = \frac{t}{E(t)} \tag{3.46}$$

where $E(t)$ is the mean of the underlying distribution. This bound is associated with the elementary renewal theorem expressing the asymptotic behavior of any RP. The theorem states that

$$\lim_{t \to \infty} \frac{M_{RP}(t)}{t} = \frac{1}{E(t)} \tag{3.47}$$

where $M_{RP}(t)$ is the CIF of an RP.

Figure 3.19 shows the upper bounds (3.40), (3.44), and (3.46) for the RP having the Weibull underlying distribution with the scale parameter equal to 1 and the shape parameter equal to 1.5.

It is easy to show that the upper bound given by (3.44) is sharper than the Barlow and Proschan upper bound on the CIF of RP (3.46) in the interval $(0, t^*)$, where t^* is given by a solution of the following equation:

---

[*] NBUE is *new better than used in expectation*. This class includes the IFR distributions.

$$\frac{-\ln\left(1-F(t)\right)}{t} = \frac{1}{E(t)} \qquad (3.48)$$

For the bounds depicted in Figure 3.19, $t^* \approx 1.227$. Thus, the simple upper bound given by (3.44) can be effectively used for short-term applications.

For the application considered, the point t^* defined as a solution of Equation (3.48) can also be used as a boundary point between the short-term and long-term random process behavior.

Figure 3.20 depicts the long-term behavior of some of the failure process models and bounds, all of them over a time interval 10 times larger than in Figure 3.19. Note again that the CIF of the NHPP with the same IFR underlying distribution $F(t)$ is the upper bound on the CIF, not only for the RP, but also (in contrast to the Barlow and Proschan upper bound (3.46)) for the GRP with repair effectiveness parameter $q < 1$.

As illustrated by Figure 3.20, closely associated with the elementary renewal theorem, the Barlow-Proschan upper bound on the CIF of the renewal process (3.46) is practically applicable for the RP observed in a long-term interval (i.e., for $t > t^*$).

Let us note that, from the standpoint of the applications considered, neither Blackwell's theorem (see Section 3.3) nor the elementary renewal theorem that we have just discussed addresses the question of how large the value of t must be in order to apply the respective limit. In a sense, for practical use, the solution of (3.48), t^*, can be used as a lower time limit, beyond which the long-term Barlow and Proschan bound (3.46) can be effectively applied.

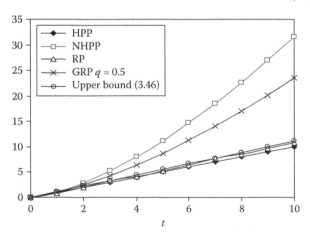

FIGURE 3.20
The cumulative intensity functions for RP, HPP, NHPP (the upper bound (3.44)), and GRP with the rejuvenation parameter $q = 0.5$, and Barlow-Proschan upper bound (3.46). The underlying distribution is the Weibull one with the scale parameter equal to 1 and the shape parameter equal to 1.5 (except for the HPP).

Now, let us discuss an analogy between (1) a positively defined random variable used as a model for the TTF of nonrepairable objects and (2) random events considered in the framework of a counting process used as a model for the failure process of repairable objects. In nonparametric TTF distribution estimation (applied to nonrepairable objects), several special classes of distributions can be used, such as the increasing/decreasing failure rate (IFR/DFR) class, the increasing/decreasing failure rate average (IFRA/DFRA) class, the new-better/worse-than-used (NBU/NWU) class, and the new-better/worse-than-used-in-expectation (NBUE/NWUE) class (Barlow and Proschan, 1996) (see Chapter 2).

Likewise, in Section 3.1 the following classes of the counting processes used for modeling failure processes of repairable objects were introduced in terms of the cumulative intensity function and its derivative (ROCOF). We call a counting process the *increasing/decreasing rate of occurrence of failures (ROCOF) process*, i.e., an IRP/DRP, if its ROCOF exists and it is an increasing/decreasing function.

The quantile bounds based on a known IFRA/DFRA cumulative distribution function were introduced in Chapter 2. Below, we introduce the similar bounds for the IRP/DRR cumulative intensity function (Kaminskiy, 2007).

It is worth mentioning that all known bounds (including those discussed above) for the renewal process (RP) and GRP are based on some assumptions about the respective underlying distribution. The bounds, which are introduced below, are *nonparametric* and *are not* expressed in terms of underlying distributions. They are applicable to any counting process with an increasing/decreasing rate of occurrence of failures.

Consider a counting process with strictly increasing cumulative intensity function $W(t) = E(N(t))$, which is the mean number of events (failures) observed in the interval $(0, t]$. Let's define the *p*th *quantile of the counting process* as the time τ_p at which $W(\tau_p) = p$, where $p > 0$. Note that, in contrast to the quantile of a random variable, the quantile of a counting process can be greater than 1. Below, we consider a counting process for which ROCOF exists and is an increasing/decreasing function. In other words, the process is IRP/DRP.

Now let's consider the IRP (DRP) counting process with *p*th quantile τ_p (that is, $W(\tau_p) = p$). Then,

$$W(t) \leq (\geq) p \frac{t}{\tau_p} \qquad \text{for } 0 \leq t < \tau_p$$

$$W(t) \geq (\leq) p \frac{t}{\tau_p} \qquad \text{for } t \geq \tau_p$$

(3.49)

Inequalities (3.49) are not difficult to prove. Consider the HPP with the same *p*th quantile as the IRP of interest (see Figure 3.21). The cumulative

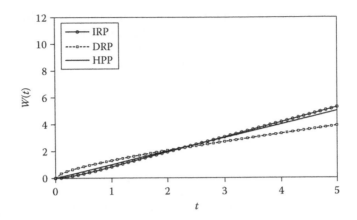

FIGURE 3.21
The CIFs for IRP (DRP), which is the renewal process with the underlying Weibull distribution
with the scale parameter equal to 1 and the shape parameter equal to 1.5 (0.5). The HPP line is
the CIF for the homogeneous Poisson process (the exponential underlying distribution with
scale parameter equal to 1).

intensity function of the HPP is linear. According to the IRP definition,
its cumulative intensity function is concave upward. Both functions pass
through the origin, and the cumulative intensity function of the IRP crosses
the cumulative intensity function of the respective homogeneous Poisson
process at τ_p. Thus, the segment of the cumulative intensity function of the
homogeneous Poisson process in the time interval $[0, \tau_p]$ is the chord joining
the point $(0, 0)$ to (τ_p, p) and lying above the graph of the cumulative intensity
function of IRP, and so (3.49) holds. For DRP, the proof is similar. The example considered below is, to an extent, similar to the example given by Barlow
and Proschan (1996) for nonrepairable objects having IFR life distribution.

EXAMPLE 3.4
A contractor is required to produce a repairable system admitting, on
average, five failures during a mission length of 10,000 h. Time to repair
is considered negligible. Assuming the contractor just meets the require-
ment, what is a conservative prediction of the expected number of failures during the first 3,000 h of the mission?

SOLUTION
Using (3.49), $t = 3{,}000$ h, $p = 5$, and $\tau_p = 10{,}000$. Noting that $t < \tau_p$, and evaluating p_t/τ_p, we find that $W(3{,}000) \leq 1.5$. Thus, given the above-described
assumptions, the contractor may conservatively claim that the system
will require on average not more than two repairs during the first 3,000 h
of its mission.

3.7 Geometric Process: Adding the Better than New Repair

One limitation of the GRP model is its inability to model a better than new restoration, which is needed for some practical applications, e.g., *reliability growth modeling* (Crow, 1982). The *geometric process* (Lam, 1988, 2009) considered below overcomes this particular drawback.

First, we have to remark that the better than new restoration can't be modeled using the notion of age. Thus we have to recall the family of *location-scale distributions*, for which the geometric process was introduced.

For a continuous random variable t the location-scale distribution is defined as having the cumulative distribution function (CDF) in the following form:

$$F(t) = F\left(\frac{t-u}{\alpha}\right) \tag{3.50}$$

where u is the location parameter and α is the scale parameter. It should be noted that practically all the continuous time-to-failure distributions (or their transformations) considered above belong to the family of location-scale distributions.

In the framework of the geometric process, the times between successive failures X_i ($i = 1, 2, \ldots$) are assumed to be independent random variables. After each ith failure, the system is restored (damaged) in such a way that its scale parameter α is changed to $\alpha(1 + q)^{i-1}$, where $-1 < q < \infty$ is the restoration (damage) parameter, so that for the time to the first failure $i = 1$, for the time between the first and second failure $i = 2$, and so on. This transformation of the scale parameter is equivalent to the one we used in Chapter 2 considering the accelerated life model.

To an extent, the geometric process, as a model, makes more physical (reliability) sense than, say, the respective NHPP model in terms of restoration assumption (i.e., the same-as-old assumption). If $q = 0$, the process coincides with the ordinary renewal process. If $q > 0$, the introduced process is obviously an improving one, and if $q < 0$, the process is aging (deteriorating). Table 3.4 shows the multiplier $(1 + q)^{i-1}$ to the scale parameter of the underlying distributions of the times between consecutive events for some values of q.

The geometric process times between successive failures X_i ($i = 1, 2, \ldots$) are distributed according to the following cumulative distribution function (CDF):

$$F_i(X_i) = F\left(\frac{X_i - u}{\alpha(1+q)^{i-1}}\right), i = 1, 2, \ldots, n \tag{3.51}$$

The respective distributions of the time to the nth failure T_n, as well as the equations for cumulative intensity functions (CIFs) and ROCOF, are difficult

TABLE 3.4

Multiplier $(1 + q)^{i-1}$ to the Scale Parameter of the
Underlying Distributions of the Times between
Successive Events for Some Values of q

Event, i	$q = 0.1$	$q = -0.1$	$q = 0.2$	$q = -0.2$
1	1.000	1.000	1.000	1.000
2	1.100	0.900	1.200	0.800
3	1.210	0.810	1.440	0.640
4	1.331	0.729	1.728	0.512
5	1.464	0.656	2.074	0.410
6	1.611	0.590	2.488	0.328

to find as closed-form expressions. The only way to find these functions is to apply a numerical approach. Below we consider two particular cases of the geometric process: the case of underlying exponential distribution and the case of underlying Weibull distribution. In both cases, the respective CIFs were obtained using the Monte Carlo simulations.

3.7.1 Geometric Process with Exponential Underlying Distribution

Figures 3.22 and 3.23 show the cumulative intensity functions (CIFs) of the geometric process with an underlying exponential distribution. It is interesting to note that in the context of the geometric process, the underlying

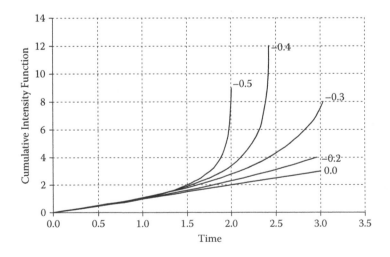

FIGURE 3.22
CIFs of the geometric process with underlying exponential distribution, scale parameter of 1, and various negative values of restoration parameter q.

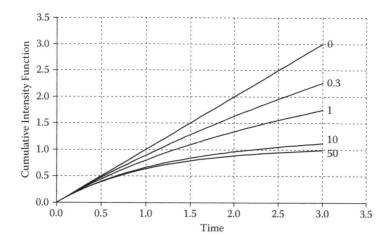

FIGURE 3.23
CIFs of the geometric process with underlying exponential distribution, scale parameter of 1, and various positive values of restoration parameter q.

exponential distribution provides a high flexibility in modeling both *improving* and *deteriorating* processes—contrary to the HPP.

3.7.2 Geometric Process with Weibull Underlying Distribution

Figures 3.24 through 3.27 show the CIFs of the geometric process with different Weibull underlying distributions (aging, i.e., increasing failure rate (IFR) and rejuvenating, i.e., DFR) and different (positive and negative) restoration parameters q.

   These figures reveal that the geometric process can be used as a very flexible model that, among other applications, might be applied to avalanche-looking processes (Figures 3.22 and 3.26), like the avalanche breakdown in solids (McKay, 1954; Tan et al., 2010). The maximum likelihood estimates for the geometric process with Weibull underlying distribution can be found in Kaminskiy and Krivtsov (2010a,b).

EXAMPLE 3.5
Consider failure times between 12 consecutive failures discussed by Basu and Rigdon (2000): 3, 6, 11, 5, 16, 9, 19, 22, 37, 23, 31, and 45. The geometric process with the underlying exponential distribution is assumed as a probabilistic model. Figure 3.28 shows the maximum likelihood estimates of the CIF. It is interesting to note that the CIF exhibits a *convexity*, contrary to *linearity*, which might be intuitively expected from a point process with the underlying exponential distribution. The exponential distribution scale parameter is estimated to be 4.781 and the restoration parameter q as 0.232.

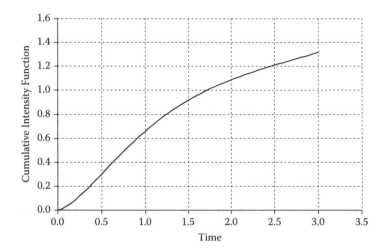

FIGURE 3.24
CIFs of the geometric process with underlying Weibull distribution, scale parameter of 1, shape parameter of 1.5, and restoration parameter q of 3.

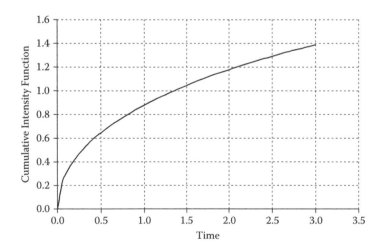

FIGURE 3.25
CIFs of the geometric process with underlying Weibull distribution, scale parameter of 1, shape parameter of 0.5, and restoration parameter q of 3.

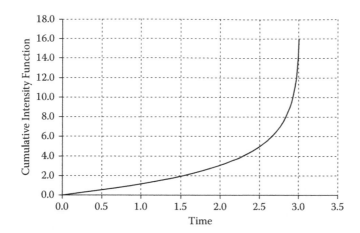

FIGURE 3.26
CIFs of the geometric process with underlying Weibull distribution, scale parameter of 1, shape parameter of 1.5, and restoration parameter q of -0.3.

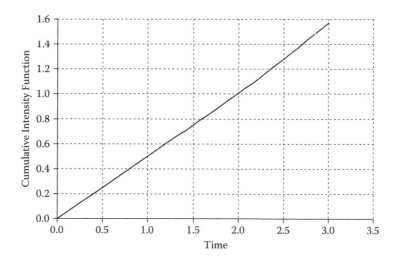

FIGURE 3.27
CIFs of the geometric process with underlying Weibull distribution, scale parameter of 1, shape parameter of 0.5, and restoration parameter of -0.3.

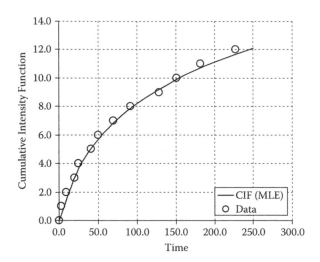

FIGURE 3.28
Geometric process with exponential underlying distribution as a model of the data set of Basu and Rigdon (2000).

3.8 Gini-Type Index for Aging/Rejuvenating Processes

In Chapter 2, we introduced the Gini-type index (Kaminskiy and Krivtsov, 2010), helping to evaluate how far a given *distribution* deviates from the exponential one. Now, we will introduce a similar index for repairable systems modeled by the *point processes*. We will see that in both cases, the index takes on the values between –1 and 1.

Consider a point process having an integrable over [0, *T*] cumulative intensity function, $\Lambda(t)$. It is assumed that the respective ROCOF exists, and it is an increasing function over the same interval [0, *T*], so that $\Lambda(t)$ is concave upward, as illustrated by Figure 3.29. Further consider the homogeneous Poisson process (HPP) with cumulative intensity function (CIF) $\Lambda_{HPP}(t) = \lambda t$ that coincides with $\Lambda(t)$ at $t = T$, i.e., $\Lambda_{HPP}(T) = \Lambda(T)$ (see Figure 3.29).

For a given time interval [0, *T*], the GT index is defined as

$$C(T) = 1 - \frac{\int_0^T \Lambda(t)dt}{0.5T\Lambda(T)} = 1 - \frac{2\int_0^T \Lambda(t)dt}{T\Lambda(T)} \tag{3.52}$$

The smaller the absolute value of the GT index, the closer the considered point process is to the HPP; clearly, for the HPP, $C(T) = 0$. The GT index satisfies the following inequality: $-1 < C(T) < 1$. It is obvious that for a point process with an increasing ROCOF, the GT index is positive, and for a

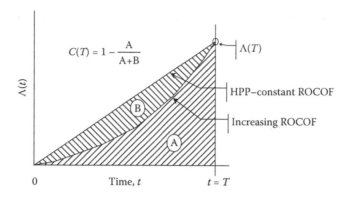

FIGURE 3.29
Graphical interpretation of the GT index for a point process with an increasing ROCOF.

point process with a decreasing ROCOF, the index is negative. One can also show that the absolute value of the GT index $C(T)$ is proportional to the mean distance between the $\Lambda(t)$ curve and the CIF of the HPP.

For the most popular nonhomogeneous Poisson process (NHPP) model—the *power law* model with the underlying Weibull distribution—the GT index is expressed in a closed form:

$$C = 1 - \frac{2}{\beta + 1} \tag{3.53}$$

where β is the shape parameter of the underlying Weibull distribution.

It is interesting that (3.53) is exactly the same as the GT index for the Weibull distribution (2.61) from Chapter 2. This is because the NHPP's CIF is formally equal to the cumulative hazard function of the underlying failure time distribution (see, e.g., Krivtsov, 2007). Some examples of the GT index for other point processes commonly used in reliability and risk analysis are given in Table 3.5.

TABLE 3.5

GT Indexes of Some Point Processes over Time Interval [0, 2]

Point Process	Shape Parameter of Underlying Weibull Distribution	Rejuvenation Parameter, q	GT Index
HPP	1	0	0
NHPP	1.1	1	0.05
NHPP	2	1	0.33
NHPP	3	1	0.50
RP	2	0	0.82
GRP	2	0.5	0.21

Notes: The GT index for RP and GRP was obtained using numerical techniques. All the underlying distributions are Weibull ones with scale parameter $\alpha = 1$.

Thus, in Sections 2.2 and 3.6, we have introduced a simple index that helps to assess the degree of aging or rejuvenation applicable to repairable objects (systems) as well as to nonrepairable objects (components). The index ranges from –1 to 1. It is negative for the class of decreasing failure rate distributions and point processes with decreasing ROCOF, and it is positive for the increasing failure rate distributions and point processes with increasing ROCOF. The index might also be found useful in hypothesis testing for the exponentiality of the TTF or failure interarrival times.

Exercises

1. Show that the upper bound given by (3.44) is sharper than the Barlow and Proschan (1996) upper bound on the CIF of RP (3.46) in the interval $(0, t^*)$, where t^* is given by a solution of the following equation:

$$\frac{-\ln(1-F(t))}{t} = \frac{1}{E(t)} \qquad (3.48)$$

2. Show that the absolute value of GT index $C(T)$ is proportional to the mean distance between the $\Lambda(t)$ curve and the CIF of the HPP.

Appendix A: Transformations of Random Variables

Below, we consider a simple case of one random variable (r.v.). Let's have a single r.v. X ($-\infty < X < \infty$) with the probability density function (PDF) $f_X(X)$ and the cumulative distribution function (CDF) $F_X(X)$. Let $g(X)$ be a monotone function transforming r.v. X into a new (dependent) r.v. $Y = g(X)$. The function $g(X)$ is assumed to have a derivative everywhere. Our problem is to find the PDF $f_Y(Y)$ of the dependent r.v. Y.

Because Y is a monotone function of X, the probability that X is less than a certain value x is the probability that Y is less than the respective value $y = g(x)$, i.e., $\Pr(X < x) = \Pr(Y < g(x))$. So, we can write

$$F_Y(y) = \Pr(Y < y) = \Pr(g(x) < y) = \Pr(X < g^{-1}(y)) = \int_{-\infty}^{g^{-1}(y)} f_X(\xi)d\xi \quad (A1.1)$$

where $F_Y(Y)$ is the CDF of Y, and $g^{-1}(y)$ is the inverse function of $g(y)$. For the sake of simplicity, we are considering only the case where the inverse function $g^{-1}(y)$ is a simple (one-to-one) function. Taking the derivative of (A1.1) gives the PDF of the dependent r.v. Y:

$$f_Y(y) = \frac{dF_Y(y)}{dy} = \frac{d\left[\int_{-\infty}^{g^{-1}(y)} f_X(\xi)d\xi\right]}{dx} \frac{dx}{dy} \quad (A1.2)$$

$$= f_X\left[g^{-1}(y)\right] \frac{d\left[g^{-1}(y)\right]}{dy}$$

Forcing the term

$$\frac{d\left[g^{-1}(y)\right]}{dy}$$

to be always positive, so that $f_Y(Y)$ is positive, results in the following relationship for a function of our r.v. X:

$$f_Y(y) = f_X\left[g^{-1}(y)\right] \left|\frac{d\left[g^{-1}(y)\right]}{dy}\right| \quad (A1.3)$$

Appendix B: Coherent Systems

Let's consider a system composed of n components. It is assumed that each component and the system itself can be only in two states—they can be either in a functioning state or in a failed state. Let's denote the functioning state by 1 and the failed state by 0, and introduce a binary indicator variable x_i related to the ith components as (Barlow and Proschan, 1975)

$$x_i = \begin{cases} 0 & \text{if component is } \textit{failed} \\ 1 & \text{if component is } \textit{functioning} \end{cases} \tag{A2.1}$$

for $i = 1, 2, ..., n$. Similarly, the binary state of the system (*functioning* or *failed*) is described by a *binary function* φ, taking on the value 0 or 1 according to the following rule:

$$\varphi = \begin{cases} 0 & \text{if the system is } \textit{failed} \\ 1 & \text{if the system is } \textit{functioning} \end{cases}$$

It is assumed, that the state of the system is completely determined by the *state vector* $\boldsymbol{x} = (x_1, x_2, ..., x_n)$, so that one can write

$$\varphi = \varphi(\boldsymbol{x}) \tag{A2.2}$$

The function $\varphi(x)$ is called the *structure function* of the system. As an example, let's consider a series system composed of n components. The system is functioning if each component functions. The respective structure function is given by

$$\varphi(\boldsymbol{x}) = \prod_{i=1}^{n} x_i \tag{A2.3}$$

More examples of the structure functions of the popular system reliability structures can be found in Barlow and Proschan (1975), Leemis (1995, 2009), and many other reliability books.

A system is *coherent* if its structure function is nondecreasing in x, and if it does not include any irrelevant components. To illustrate the nondecreasing in x structure function consider the following inequality:

$$\varphi(x_1, x_2, ..., x_{i-1}, ..., 0, x_{i+1}, ..., x_n) \leq \varphi(x_1, x_2, ..., x_{i-1}, ..., 1, x_{i+1}, ..., x_n)$$

The second part of the coherent system definition is related to the notion of irrelevant component. It is said that a component is irrelevant if its state does not affect the structure function of the system to which the component belongs. One of the most important results related to the coherent systems is the theorem stating that redundancy at the component level is more effective than redundancy at the system level (Barlow and Proschan, 1975).

Appendix C: Uniform Distribution

The probability density function of the uniform distribution is given by (see Figure A3.1):

$$f(x;a, b) = \begin{cases} \dfrac{1}{b-a} & a < x \leq b \\ 0 & \text{otherwise} \end{cases} \tag{A3.1}$$

where a and b are the distribution parameters.

The corresponding cumulative distribution function of the uniform distribution is

$$F(x;a, b) = \begin{cases} 0 & x < a \\ \dfrac{x-a}{b-a} & a < x \leq b \\ 1 & x > b \end{cases} \tag{A3.2}$$

When $a = 0$ and $b = 1$, the uniform distribution is called the standard uniform distribution. The mean of the standard uniform distribution is 0.5, and the variance is 1/12.

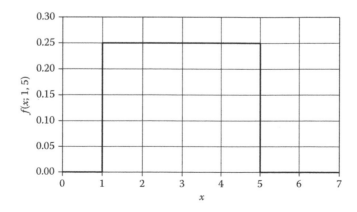

FIGURE A3.1
Probability density function of uniform distribution with parameters $a = 1$ and $b = 5$.

References and Bibliography

Asher, H. and Feingold, H., *Repairable Systems Reliability: Modeling and Inference, Misconception and Their Causes*, Marcel Dekker, New York, 1984.

Atwood, C., Parametric Estimation of Time-Dependent Failure Rates for Probabilistic Risk Assessment, *Reliability Eng. Syst. Safety*, 37, 181–194, 1992.

Barlow, R.E., and Proschan, F., *Statistical Theory of Reliability and Life Testing Probability Models*, Holt, Rinehart and Winston, New York, 1975.

Barlow, R.E., and Proschan, F., with contributions by L. Hunter, *Mathematical Theory of Reliability*, SIAM Series in Applied Mathematics, Wiley, New York, 1996.

Barlow, R.E., and Proschan, F., with contributions by L. Hunter, *Mathematical Theory of Reliability*, SIAM Series in Applied Mathematics, Wiley, New York, 1965.

Barlow, R.E., and Proschan, F., *Statistical Theory of Reliability and Life Testing: Probability Models*, To Begin With, Silver Spring, MD, 1981.

Barlow, R.E., Marshall, A.W., and Proschan, F., Properties of Probability Distributions with Monotone Hazard Rates, *Ann. Math. Stat.*, 34, 375–389, 1963.

Barlow, R.E., and Gupta, S., Distribution-Free Life Test Sampling Plans, *Technometrics*, 8, 591–603, 1966.

Barlow, R.E., Geometry of the Total Time on Test Transform, *Naval Res. Logistics Q.*, 26(3), 393–402, 1979.

Basu, A.P., and Rigdon, S.E., *Statistical Methods for the Reliability of Repairable Systems*, Wiley, New York, 2000.

Baxter, L. A. et al., On the Tabulation of Renewal Function, *Technometrics*, 24, 151–156, 1982.

Black, J.R., Electromigration—A Brief Survey and Some Recent Results, *IEEE Trans. Electron Devices*, 16(4), 338–347, 1969a.

Black, J.R., Electromigration Failure Modes in Aluminum Metallization for Semiconductor Devices, *Proc. IEEE*, 57(9), 1587–94, 1969b.

Blischke, W.R., and Murthy, D.N.P., *Warranty Cost Analysis*, Marcel Dekker, New York, 1994.

Bogdanoff, J.L., and Kozin, F., *Probabilistic Models for Cumulative Damage*, John Wiley & Sons, New York, 1985.

Bolotin, V.V., *Prediction of Service Life for Machines and Structures*, ASME Press, New York, 1989.

Bolotin, V.V., *Random Vibrations of Elastic Systems*, Martinus Nijhoff Publishers, The Hague, 2010.

Bortkiewicz, L.J., *Das Gesetz der Kleinen Zahlen*, Leipzig, Teubner, 1898.

Brown, M., and Proschan, F., Imperfect Maintenance, in *Survival Analysis*, ed. J. Crowley and R. Johnson, IMS Lecture Notes—Monograph Series, Vol. 2, Hayward, 179–188, 1982.

Castillo, E., *Extreme Value Theory in Engineering (Statistical Modeling and Decision Science)*, Academy Press, San Diego, 1988.

Chan, C.K., A Proportional Hazard Approach to SiO_2 Breakdown Voltage, *IEEE Trans. Reliability*, R-39, 147–150, 1990.

Christiansen, E.L., Hyde, J.L., and Bernhard, R.P., Space Shuttle Debris and Meteoroid Impacts, *Adv. Space Res.*, 34, 1097–1103, 2004.

Chugunov, A.I., and Horowitz, C.J., Breaking Stress of Neutron Star Crust, *Monthly Notices R. Astron. Soc. Lett.*, 407(1), L54–L58, 2010.

Cook, R.J., and Lawless, J.F., *The Statistical Analysis of Recurrent Events (Statistics for Biology and Health)*, Springer, 2010.

Cox, D.R., and Lewis, P.A., *The Statistical Analysis of Series of Events*, Methuen, London, 1966.

Cox, D.R., and Oakes, D., *Analysis of Survival Data*, Chapman & Hall, London, 1984.

Crow, E.L., and Shimizu, K. (eds.), *Lognormal Distributions (Statistics: A Series of Textbooks and Monographs)*, ed. E.L. Crow and K. Shimizu, Marcel Dekker, New York, 1988.

Crow, L.H., Reliability Analysis for Complex Repairable Systems, Reliability and Biometry, ed. F. Proschan and R.J. Serfling, SIAM, Philadelphia, 1974.

Crow, L.H., Confidence Interval Procedures for the Weibull Process with Applications to Reliability Growth, *Technometrics*, 24, 67–72, 1982.

Crow, L.H., Evaluating the Reliability of Repairable Systems, *Proceedings of Annual Reliability and Maintenance Symposium IEEE*, Orlando, FL, 1990.

Crowder, M.J., Kimber, A.C., Smith, R.L., and Sweeting, T.J., *Statistical Analysis of Reliability Data*, Chapman & Hall, London, 1991.

Daykin, C.D., Penticainen, T., and Pesonen, M., *Practical Risk Theory for Actuaries*, Chapman & Hall, London, 1995.

Dodson, J., First Lectures on Insurance, 1756, available at http://kabele.org/papers/dodsonms.pdf.

Duane, J.T., Learning Curve Approach to Reliability Monitoring, *IEEE Trans. Aerospace Electron. Syst.*, 2, 563–566, 1964.

Finkelstein, M., The Concealed Age of Distribution Function and the Problem of General Repair, *J. Stat. Planning Inference*, 65, 315–321, 1997.

Fischer, H., *A History of the Central Limit Theorem: From Classical to Modern Probability Theory*, Springer, 2010.

Fisher, L.D., and Lin, D.Y., Time-Dependent Covariates in the Cox Proportional-Hazards Regression Model, *Ann. Rev. Public Health*, 20, 145–157, 1999.

Fisher, R.A., and Tippett, L.H.C., Limiting Forms of the Frequency Distribution of the Largest and Smallest Member of a Sample, *Proc. Cambridge Phil. Soc.*, 24, 180–190, 1928.

Gail, M.H., and Gastwirth, J.L., A Scale-Free Goodness-of-Fit Test for the Exponential Distribution Based on the Gini Statistic, *Journal of Royal Statistical Society B*, 40, 3, 350–357, 1978.

Garg, A., and Kalagnanam, J., Approximations for the Renewal Functions, *IEEE Trans. Reliability*, 47(1), 66–72, 1998.

Gavrilov, L.A., and Gavrilova, N.S., Models of Systems Failure in Aging, in *Handbook of Models for Human Aging*, ed. P.M. Conn, Elsevier Academic Press, Burlington, MA, 45–68, 2006.

Gavrilov, L.A., and Gavrilova, N.S., Why We Fall Apart. Engineering's Reliability Theory Explains Human Aging, *IEEE Spectrum*, 41(9), 30–35, 2004.

Gavrilov, L.A., and Gavrilova, N.S., *The Biology of Life Span*, Harwood Academic Publishers, New York, 1991.

Gibbon, E., *History of the Decline and Fall of the Roman Empire*, Elibron Classics, Adamant Media Corporation, 2005.

Gnedenko, B.V., Sur la Distribution Limite du Terme Maximum d'Une Serie Aleatoire, *Ann. Math.*, 44, 423–453, 1943.

Gnedenko, B.V., Pavlov, I., and Ushakov, I., *Statistical Reliability Engineering*, John Wiley & Sons, New York, 1999.

Goldman, A. Ya, *Prediction of the Deformation Properties of Polymeric and Composite Materials*, American Chemical Society, Washington, DC, 1994.

Guess, F., and Proschan, F., Mean Residual Life: Theory and Applications, in *Handbook of Statistics 7: Quality Control and Reliability*, ed. P.R. Krishnaiah and C.R. Rao, Elsevier Science Publishers, Amsterdam, 1988.

Guo, H.R., Liao, H., Zhao, W., and Mettas, A., Approximations for the Renewal Functions, *IEEE Trans. Reliability*, 56(1), 40–49, 2007.

Hahn, G.J., and Shapiro, S.S., *Statistical Models in Engineering*, John Wiley & Sons, New York, 1994.

Herodian, *A History of the Roman Empire since the Death of Marcus Aurelius*, trans. E.C. Echols, Berkeley, Los Angeles, 1961.

Hoyland, A., and Rausand, M., *System Reliability Theory: Models and Statistical Methods*, John Wiley & Sons, New York, 1994.

Hoyland, A., and Rausand, M., *System Reliability Theory: Models and Statistical Methods*, 2nd ed., John Wiley & Sons, New York, 2004.

Hurtado, J.L., Joglar, F., and Modarres, M., Generalized Renewal Process: Models, Parameter Estimation and Applications to Maintenance Problems, *Int. J. Performability Eng.*, 1, 2005.

IEC International Standard, *Reliability Growth—Statistical Test and Estimation Methods*, IEC 1164, International Electrotechnical Commission, 1995.

Jiménez, P., and Villalón, R., Generalized Renewal Process as an Adaptive Probabilistic Model, *IEEE PES Transmission and Distribution Conference and Exposition Latin America*, Venezuela, 1–6, 2006.

Kahle, W., Statistical Models for the Degree of Repair in Incomplete Repair Models, *International Symposium on Stochastic Models in Reliability, Safety, Security and Logistics*, Beer Sheva, Israel, 178–181, 2005.

Kaminskiy, M.P., and Krivtsov, V.V., A Statistical Estimation of the Cost Impact from Introducing a Mileage Limit in Automobile Warranty Policy, *Inst. Math. Stat. Bull.*, 28(2), 73, 1999.

Kaminskiy, M.P., and Ushakov, I.A., Practical Reliability Concepts, in *Product Reliability, Maintainability, and Supportability*, ed. M. Pecht, CRC Press, Boca Raton, FL, 1995.

Kaminskiy, M.P., and Krivtsov, V.V., A Monte Carlo Approach to Repairable System Reliability Analysis, in *Probabilistic Safety Assessment and Management*, Springer-Verlag, New York, 1063–1068, 1998.

Kaminskiy, M.P., and Krivtsov, V.V., Generalized Renewal Process as a Model for Statistical Warranty Prediction, *Proceedings of the Annual Reliability and Maintainability Symposium*, 276–279, January 2000a.

Kaminskiy, M.P., and Krivtsov, V.V., A Monte Carlo Approach to Estimation of G-Renewal Process in Warranty Data Analysis, *Proceedings 2nd International Conference on Mathematical Methods in Reliability*, Bordeaux, France, June 2000b, pp. 583–586.

Kaminskiy, M.P., Simple Bounds on Cumulative Intensity Functions of Renewal and G-Renewal Processes with Increasing Failure Rate Underlying Distributions, *Risk Anal. Int. J.*, 24(4), 1035–1039, 2004.

Kaminskiy, M.P., Simple Bounds for Counting Processes with Monotone Rate of Occurrence of Failures, *Reliability Syst. Safety*, 92, 566–568, 2007.

Kaminskiy, M.P., and Krivtsov, V.V., A Gini-Type Index for Aging/Rejuvenating Objects, in *Mathematical and Statistical Models and Methods in Reliability: Applications to Medicine, Finance, and Quality Control*, ed. V.V. Rykov, N. Balakrishnan, and M.S. Nikulin, Springer, Boston, 2010a.

Kaminskiy, M.P., and Krivtsov, V.V., G1-Renewal Process as Repairable System Model, *Stat. Methodol.*, eprint arXiv:1006.3718, 2010b.

Kapur, K.C., and Lamberson, L.R., *Reliability in Engineering Design*, Wiley, New York, 1977.

Kijima, M., and Sumita, N., A Useful Generalization of Renewal Theory: Counting Process Governed by Non-negative Markovian Increments, *J. Appl. Probability*, 23, 71–88, 1986.

Kijima, M., Morimura, H., and Suzuki, Y., Periodical Replacement Problem without Assuming Minimal Repair, *Eur. J. Oper. Res.*, 37, 194–203, 1988.

Klinger, D.J., Nakada, Y., and Menendez, M.A., *AT & T Reliability Manual*, Chapman & Hall, London, 1990.

Krivtsov, V.V., Practical Extensions to NHPP Applications in Repairable System Reliability Analysis, *Reliability Eng. Syst. Safety*, 92(5), 560–562, 2007.

Krivtsov, V.V., and Frankstein, M., Nonparametric Estimation of Marginal Failure Distributions from Dually Censored Automotive Data, *Proceedings of Annual Reliability and Maintainability Symposium*, Los Angeles, CA, January 2004, pp. 86–89.

Kotz, S., and Nadarajah, S., *Extreme Value Distributions: Theory and Applications*, Imperial College Press, London, 2000.

Lai, Ch.-D., and Min, X., *Stochastic Ageing and Dependence for Reliability*, Springer, 2006.

Lam, Y., Geometric Process and Replacement Problem, *Acta Math. Appl. Synica*, 4, 366–377, 1988.

Lam, Y., A Geometric Process D-Shock Maintenance Model, *IEEE Trans. Reliability*, 58(2), 389–396, 2009.

Lawless, J.F., *Statistical Models and Methods for Lifetime Data*, John Wiley & Son, New York, 1982.

Lawless, J.F., *Statistical Models and Methods for Lifetime Data*, 2nd ed., John Wiley & Son, New York, 2002.

Lawless, J.F., Hu, J., and Cao, J., Methods for the Estimation of Failure Distributions and Rates from Automobile Warranty Data, *Lifetime Data Anal.*, 1, 227–240, 1995.

Laz, P.J., and Hillberry, B.M., Fatigue Life Prediction from Inclusion Initiated Cracks, *Int. J. Fatigue*, 20(4), 263–270, 1998.

Lee, S., Wilson, J.R., and Crawford, M.M., Modeling and Simulation of a Nonhomogeneous Poisson Process Having Cyclic Behavior, *Commun. Stat. Simul. Comput.*, 20(2–3), 777–809, 1991.

Leemis, L., *Reliability: Models and Statistical Methods*, Prentice-Hall, Englewood Cliffs, NJ, 1995.

Leemis, L., *Reliability: Models and Statistical Methods*, 2nd ed., Prentice-Hall, Englewood Cliffs, NJ, 2009.

Levin, G., and Christiansen, E.L., The Space Shuttle Program Pre-Flight Meteoroid and Orbital Debris Risk/Damage Predictions and Post Flight Damage Assessments, ESA SP-393, in *Proceedings of the Second European Conference on Space Debris*, 633–636, 1997.

Lindqvist, B. H., Statistical Modeling and Analysis of Repairable Systems, in *Statistical and Probabilistic Models in Reliability*, Birkhauser, Berlin, 3–25, 1999.

List of Countries by Income Inequality, 2011, http://en.wikipedia.org/wiki/List_of_countries_by_income_equality.

Lloyd, D., and Lipow, M., *Reliability: Management, Methods, and Mathematics*, Prentice-Hall, Englewood Cliffs, NJ, 1962.

Lu, M.W., Automotive Reliability Prediction Based on Early Field Failure Warranty Data, *Qual. Reliability Eng. Int.*, 14(2), 103–108, 1998.

McKay, K.G., Avalanche Breakdown in Silicon, *Phys. Rev.*, 94(4), 77–884, 1954.

Mann, N.R.E., Schafer, R.E., and Singpurwalla, N.D., *Methods for Statistical Analysis of Reliability and Life Data*, John Wiley & Son, New York, 1974.

Meeker, W., and Escobar, L., *Statistical Methods for Reliability Data*, John Wiley & Sons, New York, 1998.

MIL-HDBK-781A, *Military Handbook: Reliability Test Methods, Plans, and Environments for Engineering, Development Qualification, and Production*, 1996.

Modarres, M., Kaminskiy, M., and Krivtsov, V., *Reliability Engineering and Risk Analysis: A Practical Guide*, 2nd ed., CRC Press, Boca Raton, FL, 2010.

Nelson, W., Analysis of Performance Degradation Data from Accelerated Tests, *IEEE Trans. Reliability*, R30, 149–155, 1981.

Nelson, W., *Applied Life Data Analysis*, John Wiley & Sons, New York, 1982 and 2003 (paperback edition).

Nelson, W., *Accelerated Testing: Statistical Models, Test Plans, and Data Analysis*, John Wiley & Sons, New York, 1990 and 2004 (paperback edition).

Nelson, W., *Recurrent Events Data Analysis for Product Repairs, Disease Recurrences, and Other Applications*, SIAM, ASA, Philadelphia, 2003.

Pearson, K., Notes on the History of Correlation, *Biometrika*, 13(1), 25–45, 1920.

Preece, D., Ross, G., and Kirby, P., Bortkewitsch's Horse-Kicks and the Generalized Linear Model, *J. R. Stat. Soc. D*, 37(3), 313–318, 1988.

Provan, J.W., Probabilistic Approaches to the Material-Related Reliability of Fracture-Sensitive Structures, in *Probabilistic Fracture Mechanics and Reliability*, ed. J.W. Provan, Martinus Nijhoff Publishers, Dordrecht, 1987.

Regel, A., Slutsker, A.I., and Tomashevsky, E.E., *Kinetic Nature of Solids Strength*, Nauka, Moscow, 1974.

Rigdon, S., and Basu, A., *Statistical Methods for the Reliability of Repairable Systems*, John Wiley & Sons, New York, 2000.

Sen, A., *On Economic Inequality*, Clarendon Press, Oxford, 1997.

Smith, W., and Leadbetter, M., On the Renewal Function for the Weibull Distribution, *Technometrics*, 5, 243–302, 1963.

Sobczyk, K., and Spencer, B.F., Jr., *Random Fatigue: From Data to Theory*, Academic Press, New York, 1992.

Tan, L.J.J., Ong, D.S.G., Ng, J.S., Tan, C.H., Jones, S.K., Qian, Y., and David, J.P.R., Temperature Dependence of Avalanche Breakdown in InP and InAlAs, *IEEE J. Quantum Electron.*, 46(8), 1153–1157, 2010.

Thompson, W.A., Jr., On the Foundation of Reliability, *Technometrics*, 23, 1–13, 1981.

Veber, B., Nagode, M., and Fajdiga, M., Generalized Renewal Process for Repairable Systems Based on Finite Weibull Mixture, *Reliability Eng. Syst. Safety*, 93, 1461–1472, 2008.

Vesely, W., Incorporating Aging Effects into Probabilistic Risk Analysis Using a Taylor Expansion Approach, *Reliability Eng. Syst. Safety*, 32, 315–337, 1991.

Wellesley, K., *The Long Year, A.D. 69*, Westview Press, Boulder, CO, 1976.

White, J., Weibull Renewal Analysis, *Proceedings of 3rd SAE Annual Reliability and Maintainability Conference*, New York , 639–657, 1964.

Wu, E.Y., and Suné, J., Power-Law Voltage Acceleration: A Key Element for Ultra-Thin Gate Oxide Reliability, *Microelectronics Reliability*, 45, 1809–1834, 2005.

Zhang, F., and Jardine, A.K.S., Optimal Maintenance Models with Minima Repairs, Periodic Overhaul, and Complete Renewals, *IIE Trans.*, 1109–1119, 1998.

Index

A

Accelerated life (AL) models, 45, 53–56. *See also* time-to-failure distributions
  cumulative damage models and, 50–51
  percentile, 50
  time transformations, 47–50
  time-dependent explanatory variables in, 58–65
  time-dependent stress, 65–68
Acceleration function, 46. *See also* Time transformation function
Activation energy, 54, 56
Actuarial business, use of Bathtub (U-shaped) failure rate function in, 36
Age, 1
  real, 100
  virtual, 100
Aging, 15, 71
  repairable objects, 83
Aging distributions. *See also* IFR distributions
  bounds and inequalities for, 39–40
  classes of, 33–36
  damage accumulation models resulting in, 36–39
  Gini-type index for, 41–45
Aging processes, Gini-type index for, 118–120
Alternating renewal process, 81
Arrhenius model, 54
Arrhenius reaction rate model, 55
As-good-as-new repair. *See* RP model
As-good-as-old repair model. *See* NHPP
Average failure rate, 3, 5

B

B_{10} bearing life, 5
Barlow and Proschan bound, 109–110

Bathtub (U-shaped) failure rate function, 36
  human mortality, 75
Bernoulli trials, 8
Better-than-new repair, geometric process, 113–115
Binomial distribution, 8–9
Biomedical life data analysis, use of proportional hazards model in, 52
Black's equation, 55
Blackwell's theorem, 91
Bortkiewicz' data set, 12
Bounds
  known mean, 40
  known quantile, 39
  RP and GRP models, 107–108

C

Case studies
  cumulative failure rate, 16
  fatigue crack modeling, 27
CDF, 1–2
  competing risks model, 69
  exponential distribution of, 15
  gamma distribution, 18
  geometric process, 113
  GRP models, 101
  Kijima models, 102
  location-scale distributions, 113
  log-linear model, 95
  lognormal distribution, 25–26
  mixture distribution model, 71–72
  NHPP, 92
  normal (Gaussian) distribution, 22
  proportional hazards model, 51
  underlying distribution, 88
  Weibull distribution, 29
Central limit theorem, 23
Characteristic life of the Weibull distribution, 29
Chemical reaction rate model, 55

Printed and bound by CPI Group (UK) Ltd, Croydon, CR0 4YY

18/10/2024

01776261-0020